Figures

A PRIMER OF
OILWELL DRILLING

A Basic Text of Oil and Gas Drilling

Sixth Edition

by Ron Baker

published by

PETROLEUM EXTENSION SERVICE
The University of Texas at Austin
Continuing & Extended Education
Austin, Texas

in cooperation with

INTERNATIONAL ASSOCIATION
OF DRILLING CONTRACTORS
Houston, Texas

2001

Library of Congress Cataloging-in-Publication Data

Baker, Ron, 1940–
A primer of oilwell drilling: a basic text of oil and gas drilling / by Ron Baker. — 6th ed.
p. cm.
ISBN 0-88698-194-8 (alk. paper)
1. Oil well drilling. I. University of Texas at Austin. Petroleum Extension Service. II. Title.
TN871.2B3162001
622'.3382—dc21
00-011902 CIP

First edition published 1951. Fifth edition 1994
Sixth edition 2001
Printed in the United States of America

Catalog No. 2.00060
ISBN 0-88698-194-8

Contents

Tables

Preface

Petroleum Extension Service (PETEX) published the first edition of *A Primer of Oilwell Drilling* in 1951. The book's section on cable-tool drilling was almost as large as the part devoted to rotary drilling, and it spent as much ink on steam power as it did on internal combustion engines. Later editions, of course, evolved with the industry; thus, the third edition (released in the early 1970s) did not so much as mention cable tools and steam power.

PETEX generally releases new editions of the Primer when changes in drilling techniques and equipment are significant enough to warrant new versions. In two cases, however, changes to the manual were relatively minor. Therefore, PETEX released them as revisions rather than new editions. Consequently, this sixth edition of the Primer is, in reality, the eighth new version of the book since its initial release in the 1950s. To say that the drilling industry has changed since then is, of course, an understatement. This new edition reflects those changes; however, it also acknowledges old techniques and philosophy and their influence on the modern drilling industry.

Regardless of how the industry has changed, the book's purpose has not: namely, to clearly explain drilling to nontechnical persons. The Primer is just that—a first reader of the oilwell drilling business. Although it is written primarily for adults, junior and senior high school students should also find it informative.

It is important to acknowledge the contributions the drilling industry has made to this manual. Tom Thomas, Transocean Sedco Forex's Modular Training Project Manager, must be particularly recognized. Enduring numerous interruptions, he gave many hours of his time and expertise to answering questions and providing sources of information. He and his company also provided encouragement and access to offshore locations without which this book could not have been rewritten. Mike Stevens, Health, Safety, and Environmental Manager at Nabors Drilling USA, Inc. arranged for many visits to Nabors rigs to obtain photographs. Crews on every Nabors rig visited were courteous, helpful, and patient. Nabors' help in providing photos is inestimable. The same is true of Helmerich and Payne International Drilling Company and Rowan Companies, Inc., whose office and field personnel assisted every step of the way. Special thanks also go to Ken Fischer of IADC, who read the manuscript and gave many valuable suggestions for improving it.

Making a good-looking book out of typed manuscript and a collection of photos and sketches is a hard job. Fortunately, the PETEX publications staff was more than up to the task. Their superlative work in editing, rewriting, drawing, photographing, laying out, and proofreading must be recognized, for without such efforts, this book could not exist in its present form.

In spite of the assistance PETEX got in writing and illustrating this primer, PETEX is solely responsible for its contents. Keep in mind, too, that while every effort was made to ensure accuracy, this manual is intended only as a training aid, and nothing in it should be considered approval or disapproval of any specific product or practice.

Ron Baker, Director
Petroleum Extension Service

Units of Measurement

Throughout the world, two systems of measurement dominate: the English system and the metric system. Today, the United States is almost the only country that employs the English system.

The English system uses the pound as the unit of weight, the foot as the unit of length, and the gallon as the unit of capacity. In the English system, for example, 1 foot equals 12 inches, 1 yard equals 36 inches, and 1 mile equals 5,280 feet or 1,760 yards.

The metric system uses the gram as the unit of weight, the metre as the unit of length, and the litre as the unit of capacity. In the metric system, for example, 1 metre equals 10 decimetres, 100 centimetres, or 1,000 millimetres. A kilometre equals 1,000 metres. The metric system, unlike the English system, uses a base of 10; thus, it is easy to convert from one unit to another. To convert from one unit to another in the English system, you must memorize or look up the values.

In the late 1970s, the Eleventh General Conference on Weights and Measures described and adopted the Système International (SI) d'Unités. Conference participants based the SI system on the metric system and designed it as an international standard of measurement.

A Primer of Oilwell Drilling gives both English and SI units. And because the SI system employs the British spelling of many of the terms, the book follows those spelling rules as well. The unit of length, for example, is *metre*, not *meter*. (Note, however, that the unit of weight is *gram*, not *gramme*.)

To aid U.S. readers in making and understanding the conversion to the SI system, we include the following table.

English-Units-to-SI-Units Conversion Factors

Quantity or Property	English Units	Multiply English Units By	To Obtain These SI Units
Length, depth, or height	inches (in.)	25.4	millimetres (mm)
		2.54	centimetres (cm)
	feet (ft)	0.3048	metres (m)
	yards (yd)	0.9144	metres (m)
	miles (mi)	1609.344	metres (m)
		1.61	kilometres (km)
Hole and pipe diameters, bit size	inches (in.)	25.4	millimetres (mm)
Drilling rate	feet per hour (ft/h)	0.3048	metres per hour (m/h)
Weight on bit	pounds (lb)	0.445	decanewtons (dN)
Nozzle size	32nds of an inch	0.8	millimetres (mm)
Volume	barrels (bbl)	0.159	cubic metres (m^3)
		159	litres (L)
	gallons per stroke (gal/stroke)	0.00379	cubic metres per stroke (m^3/stroke)
	ounces (oz)	29.57	millilitres (mL)
	cubic inches (in.^{3})	16.387	cubic centimetres (cm^3)
	cubic feet (ft^3)	28.3169	litres (L)
		0.0283	cubic metres (m^3)
	quarts (qt)	0.9464	litres (L)
	gallons (gal)	3.7854	litres (L)
	gallons (gal)	0.00379	cubic metres (m^3)
	pounds per barrel (lb/bbl)	2.895	kilograms per cubic metre (kg/m^3)
	barrels per ton (lb/tn)	0.175	cubic metres per tonne (m^3/t)
Pump output and flow rate	gallons per minute (gpm)	0.00379	cubic metres per minute (m^3/min)
	gallons per hour (gph)	0.00379	cubic metres per hour (m^3/h)
	barrels per stroke (bbl/stroke)	0.159	cubic metres per stroke (m^3/stroke)
	barrels per minute (bbl/min)	0.159	cubic metres per minute (m^3/min)
Pressure	pounds per square inch (psi)	6.895	kilopascals (kPa)
		0.006895	megapascals (MPa)
Temperature	degrees Fahrenheit (°F)	$\frac{°F - 32}{1.8}$	degrees Celsius (°C)
Thermal gradient	1°F per 60 feet	—	1°C per 33 metres
Mass (weight)	ounces (oz)	28.35	grams (g)
	pounds (lb)	453.59	grams (g)
		0.4536	kilograms (kg)
	tons (tn)	0.9072	tonnes (t)
	pounds per foot (lb/ft)	1.488	kilograms per metre (kg/m)
Mud weight	pounds per gallon (ppg)	119.82	kilograms per cubic metre (kg/m^3)
	pounds per cubic foot (lb/ft^3)	16.0	kilograms per cubic metre (kg/m^3)
Pressure gradient	pounds per square inch per foot (psi/ft)	22.621	kilopascals per metre (kPa/m)
Funnel viscosity	seconds per quart (s/qt)	1.057	seconds per litre (s/L)
Yield point	pounds per 100 square feet (lb/100 ft^2)	0.48	pascals (Pa)
Gel strength	pounds per 100 square feet (lb/100 ft^2)	0.48	pascals (Pa)
Filter cake thickness	32nds of an inch	0.8	millimetres (mm)
Power	horsepower (hp)	0.7	kilowatts (kW)
Area	square inches (in.^{2})	6.45	square centimetres (cm^2)
	square feet (ft^2)	0.0929	square metres (m^2)
	square yards (yd^2)	0.8361	square metres (m^2)
	square miles (mi^2)	2.59	square kilometres (km^2)
	acre (ac)	0.40	hectare (ha)
Drilling line wear	ton-miles (tn•mi)	14.317	megajoules (MJ)
		1.459	tonne-kilometres (t•km)
Torque	foot-pounds (ft•lb)	1.3558	newton metres (N•m)

Introduction

I

If you are interested in oilwell drilling, a good way to learn about it is to visit a drilling rig. A first-time visit can be educational as well as confusing. Most drilling rigs are large and noisy and, at times, the people who work on them perform actions that don't make much sense to an uninitiated observer. A drilling rig has many pieces of equipment and most of it is huge (fig. 1). But a rig has only one purpose: to drill a hole in the ground. Although the rig itself is big, the hole it drills is usually not very big—usually less than a foot (30 centimetres) in diameter by the time it reaches final depth. The skinny hole it drills, however, can be deep: often thousands of feet or hundreds of metres. The hole's purpose is to tap an oil and gas reservoir, which more often than not lies buried deeply in the earth.

Figure 1. A drilling rig is big.

Figure 2. Sometimes, a rig can be seen from a great distance.

Although rigs operate both on land and sea—"offshore" is the oilfield term—a land rig is best for a first visit. In most cases, land rigs are easier to get to because you can drive to them. Getting to offshore rigs is more complicated, because they often work many miles (kilometres) from land and you need a boat or a helicopter to reach them.

When driving to a land rig, you'll probably see part of it long before you actually arrive at the site, especially if the terrain is not too hilly or wooded (fig. 2). One of the most distinctive parts of a drilling rig is its tall, strong structural tower called a "mast" or a "derrick" (fig. 3). Masts and derricks are tall and strong. They are strong because they have to support the great weight of the drilling tools, which can weigh many tons (tonnes).

Figure 3. The mast on this rig is 147 feet (45 metres) tall.

Rig masts and derricks are tall because they have to accommodate long lengths of pipe the rig crew raises into it during the drilling process. A mast or derrick can be as high as a 16-story building—about 200 feet (60 metres) tall. Most, however, are closer to 140 feet (43 metres) high. Even so, in flat country, a structure as lofty as a 16-story building is conspicuous.

Upon arriving at the rig, the first step is to check in with the boss. He or she is probably in a mobile home or a portable building on the site that serves as an office and living quarters. The rig boss may have the intriguing title of "toolpusher"; or, rig workers may call him or her the "rig superintendent," or the "rig manager." (Currently, most toolpushers, or rig superintendents, are men; but that's changing.) Toolpusher is the traditional term for the rig boss. It probably originated from the fondness rig workers have of calling practically every inanimate thing on a rig a tool. Thus, one who bossed the personnel using the tools also pushed the tools, in a symbolic, if not actual, sense. Nowadays, the drilling industry leans towards the term rig superintendent or rig manager for the person in charge, but you'll still hear rig hands call him or her the toolpusher (or, in Canada, the "toolpush").

Now don your hard hat, which is a very tough plastic cap with a brim to protect your head. Also, put on your steel-capped boots, which keep your toes from being crushed, and your safety glasses to safeguard your eyes. This style of dress is *de rigueur* for everyone. Whether working on a rig or merely visiting it, everyone must wear personal protective equipment, or PPE for short (fig. 4). Rig workers also wear gloves to protect their hands and you may want to wear a pair, too.

Figure 4. Personal protective equipment (PPE) includes hard hats and safety glasses.

Figure 5. Steel stairs with handrails lead up to the rig floor.

With protective gear on and the rig superintendent's permission, let's go up to the rig floor. The floor is the main work area of the rig and it usually rests on a strong foundation, a *substructure*, which raises it above ground level. Accordingly, we have to walk up a set of steel stairs (fig. 5). Keep a hand on the handrail as you walk up and don't hurry. It could be a 40-foot (12-metre) climb. Once on the floor, stop for a minute to catch your breath and take a good look around the floor. You may see the crew handling several lengths, or *joints*, of drill pipe, the steel tubes that put the *bit* (the hole-boring device) on the bottom of the hole. On the other hand, the rig may be drilling, or "making hole," as they sometimes say. If it's drilling, from time to time you may hear the distinctive and loud squawk or squeal of the drawworks brake as it slacks off the drilling line to allow the bit to drill ahead. The *drawworks* is a large, powerful hoist that, among other things, regulates the weight the drill string puts on the bit (fig. 6). A loud screech comes every time the friction brake bands ease their grip on the steel hubs of the drawworks drum to apply weight. It's loud, but it's music to the ears of the rig owner because it usually means the bit is drilling ahead without problems.

Regardless of what's occurring on the rig floor, take time to observe, for you're standing in a place that is vital to the oil and gas industry. Certainly, many operations besides drilling are involved in getting crude oil and natural gas out of the ground and into forms we can use, such as gasoline and heating fuel. However, without a drilled well—a hole in the ground—oil companies could not obtain oil and gas, or petroleum, at all.

At this point, you may not know what the equipment is for or what the personnel are doing, but don't be troubled. This book will identify most of the people and tools it takes to drill, and will give you a better appreciation of oilwell drilling. Before launching into equipment and processes, however, let's cover a little drilling history.

Figure 6. The drawworks is a powerful hoist.

History

2

The story of oilwell drilling in the United States begins in the mid-1800s, at the dawn of the industrial revolution. It was a time when people were beginning to need something better than candles to work and read by. Responding to the demand for reliable lighting, companies began making oil lamps that were brighter than candles, lasted longer, and were not easily blown out by an errant breeze.

One of the best oils to burn in these lamps was sperm-whale oil. Sperm oil was clear, nearly odorless, light in weight, and burned with little smoke. Virtually everyone preferred whale oil, but by the mid-1800s, it was so scarce that only the wealthy could afford it. The New England whalers had all but hunted their quarry to extinction. Thus, the time was ripe for an inexpensive lamp oil to replace whale oil. At the same time, steam-powered machines that required good-quality lubricants were becoming common.

About this time—1854—a New York attorney named George Bissell received a sample of an unusual liquid from a professor at Dartmouth College. Bissell and the professor had met previously and had discovered a mutual interest in finding a whale-oil substitute. The professor wanted Bissell's opinion of the liquid's value as a lamp oil and lubricant. The sample had been collected near a creek that flowed through the woods of Crawford and Venango counties in northwestern Pennsylvania. Besides water, the creek also carried an odorous, dark-colored substance that burned and, when applied to machinery, was a good lubricant. The substance was, of course, oil. Because it flowed out of the rocky terrain in and near the creek, people called it "rock oil." Indeed, so much oil flowed into the stream that settlers named it Oil Creek (fig. 7).

Figure 7. Oil Creek as it looks today

The sample came from land next to the creek just southeast of the town of Titusville, where the oil seeped from the rocks in the form of a spring.

THE DRAKE WELL, 1850s

After examining the oil sample, Bissell was convinced that refined rock oil would burn as cleanly and safely as any of the oils available at the time, including whale oil. He also believed that it would be a good lubricant. Bissell thus began raising money to collect the oil from the Titusville spring and to market it for illumination and lubrication. It was a difficult proposition; after a false start or two and much wheeling and dealing, Bissell, a Connecticut banker named James M. Townsend, and others formed what ultimately became the Seneca Oil Company, in New Haven, Connecticut.

One problem the company faced was how best to produce the oil from the land. The company directors knew that it was not efficient to simply let the oil flow out of the rock and scoop it from the ground. Others who had collected oil in this manner obtained merely a gallon (a few litres) or two a day. Seneca Oil's purpose was to produce large amounts of oil and market it in the populous northeastern U.S. Somebody in the company—no one knows who—came up with the idea of drilling a well to tap the oil. Drilling was not a new concept, for people had been drilling saltwater wells in the Titusville area for years. Interestingly, many of these saltwater wells also produced oil, which the salt drillers considered a nuisance because it contaminated the salt.

Another issue facing the fledgling oil company was the need to hire someone to oversee the drilling project in Titusville. Eventually, board member Townsend met and hired Edwin L. Drake to represent Seneca's interests at the Oil Creek site. At the time, Drake was an unemployed railroad conductor, but he had two things going for him. First, because he was out of work, he had plenty of time to devote to the project. Second, Drake had a railroad pass, which allowed him free travel to Pennsylvania. As a final touch, Townsend gave Drake the rank of honorary colonel, which sounded considerably more prestigious than just plain mister. With that, Colonel Drake went to Titusville.

By the spring of 1859, Drake employed William A. Smith to be his well driller. Smith, a blacksmith and an experienced brine-well driller, was known to most everyone as Uncle Billy. He showed up at the well site in Titusville with his sons as helpers and his daughter as camp cook. One of the first things Drake and Uncle Billy did was drive a length of hollow steel pipe through the soft surface soil until it reached bedrock. If they had not used this pipe, this steel *casing*, the loose topsoil would have caved into any hole they tried to drill. (To this day, drillers still begin oilwells by casing the top of the hole.) Drake and Smith then built the drilling rig (fig. 8), ran the drilling tools inside the casing, and drilled the rock.

Figure 8. A replica of the original Drake well; it was built in 1945 on the original site of the well. The area is now a state park.

By Saturday, August 26, 1859, Drake and Smith had drilled the hole to a depth of about 69 feet (21 metres). Near the end of the day, Smith noted that the bit suddenly dropped 6 inches (15 centimetres). It was near quitting time, so he shut the operation down, figuring he and the boys would continue drilling the following Monday. On Sunday, which in those days was a well driller's holiday, Smith decided to check on the well. He looked into the top of the casing and found the hole full of oil. Overnight, oil from a formation some 69½ feet (21.2 metres) below the surface had flowed into the well casing and filled it to the top. The well's being full of oil signaled success. No one knows for sure how much oil it produced, but it was probably around 800 to 1,200 gallons (about 3,000 to 4,800 litres) per day, which far outstripped the gallon or two that could be collected off the ground. Regardless of how much oil the well actually produced, it demonstrated that a drilled well could yield ample amounts of oil.

As far as we know, Drake's was the first well in the United States drilled for the sole purpose of finding and producing oil. News of the accomplishment spread rapidly and, because a ready market existed for refined rock oil, dozens of new rigs sprang up in the area to take advantage of the demand for it. Saltwater drillers formerly reluctant to drill oilwells changed their bias, and the first oil boom in the U.S. was underway. Refined rock oil soon became the primary lamp oil. And, as machines became more common, refined rock oil became a much sought after lubricant. Colonel Drake's well in Titusville marked the beginning of the petroleum era in the United States.

CALIFORNIA, LATE 1800s

Reports of drilling for oil in Pennsylvania soon reached all parts of the U.S., Canada, and abroad. Interest in oilwell drilling was particularly high in California, where the population was rapidly growing. After prospectors found gold at Sutter's Mill in 1849, immigrants flooded into California. Unlike the northeastern U.S., which had plenty of coal for heating and for firing boilers and other machinery, California had none. Luckily, many oil and gas seeps, similar to those in Pennsylvania, occurred in California. Therefore, as word of Drake's successful drilling venture spread, enterprising Californians applied the technology to their fields. The first successful well was drilled in 1866. It was 550 feet (168 metres) deep and produced 15 to 20 barrels (about 2 to 3 cubic metres) a day. It was considered a great success and prompted the drilling of many more wells. Oil and gas production provided much of California's energy.

THE LUCAS WELL, 1901

Before long, almost everyone in the U.S. came to depend on oil as a plentiful and inexpensive source of energy. Individuals and companies were drilling wells all over the country. Virtually anywhere entrepreneurs could erect a rig, they were drilling an oilwell. Texas was no exception.

The area around Beaumont, Texas is flat, coastal plain country. When something interrupts the flatness, people tend to notice. Consequently, practically everyone in late nineteenth-century Beaumont knew about Big Hill. Big Hill, whose formal name was Spindletop, was a dome rising about 15 feet (4.5 metres) above the surrounding plain. Enough gas seeped out of the dome that a lighted match easily ignited it.

One person particularly fascinated by Spindletop was Patillo Higgins, a self-taught geologist who lived in the region. He was convinced that oil and gas lay below Spindletop about 1,000 feet (300 metres) deep. Around 1890, Higgins obtained land on top of the dome and, with several financial partners, drilled two unsuccessful wells. The problem was that at about 350 feet (100 metres), the bit encountered a thick sand formation that the drillers called "running quicksand."

The sand was so loose it caved into the drilled hole to make further drilling impossible. Drillers ran casing, just as Drake had, attempting to combat the cave-in. The formation was so bad, however, that it crushed the casing. Discouraged, but still certain that oil lay below Spindletop, Higgins put out the word that he would lease the property to anyone willing to drill a 1,000-foot (300-metre) test well.

Ultimately, an Austrian mining engineer answered Higgins's call. Named Anthony Lucas, the engineer visited Spindletop and agreed with Higgins that the hill was a salt dome surrounded by geologic formations that trapped oil and gas. After another frustrating and costly failure, Lucas finally *spudded* (began drilling) a new well at Spindletop on October 27, 1900. He hired the Hamil brothers of Corsicana, Texas to drill the well. Aware that the running quicksand would cause trouble, the Hamils paid close attention to the mix of their *drilling fluid*. Drilling fluid is a liquid or a gas concoction that, when employed on the type of rig the Hamils used, goes down the hole, picks up the rock cuttings made by the bit, and carries the cuttings up to the surface for disposal. The type of rigs Drake and the early California drillers used did not require drilling fluid, which, as you will learn soon, all but doomed such rigs to extinction.

At Spindletop, the Hamils used water as a drilling fluid. They hand dug a pit in the ground next to the rig, filled it with water, and pumped the water into the well as they drilled it. The Hamils knew from their earlier drilling experiences, however, that clear water alone wouldn't do the job: they needed to muddy it up. They were aware that the tiny solid particles of clay in the muddy water would stick to the sides of the hole. The particles formed a thin, but strong sheath—a *wall cake*—on the sides of the hole, much like plaster on the walls of room. The wall cake stabilized the sand and kept it from caving in (fig. 9). Legend has it that the Hamils ran cattle through the earthen pit to stir up the clay and muddy the water. Whatever they did to make mud, it worked and they successfully drilled through the troublesome sand.

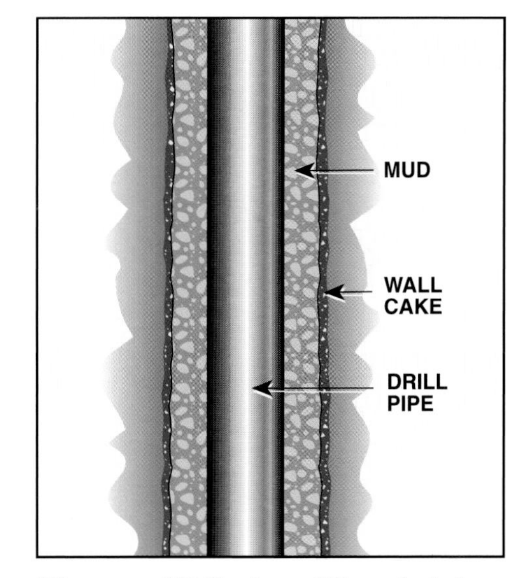

Figure 9. Wall cake stabilizes the hole.

Figure 10. The Lucas well is estimated to have flowed about 2 million gallons (8,000 cubic metres) of oil per day.

So it was that by January 1901 the new well reached about 1,000 feet (300 metres). On January 10, the drilling crew began lowering a new bit to the bottom of the hole. Suddenly, drilling mud spewed out of the well. A geyser of oil soon followed it. It gushed 200 feet (60 metres) above the 60-foot-high (18-metre-high) derrick (fig. 10). As Lucas watched the gusher from a safe distance, he estimated that it flowed at least 2 million gallons (nearly 8,000 cubic metres) of oil per day. In oilfield terms, that's about 50,000 barrels of oil per day. (One barrel of oil equals 42 U.S. gallons.) That's a lot of oil.

Thus, Spindletop's first claim to fame was that it flowed absolutely unheard of amounts of oil. Before Spindletop, a big producer flowed 2,000 barrels (320 cubic metres) per day. The Lucas well produced 25 times that amount. Spindletop's second claim to fame was that it showed the effectiveness of a type of rig, which, before Spindletop, drillers had not used much.

The Hamil's equipment was a *rotary drilling rig*; most drillers used *cable-tool rigs*. Unlike cable-tool rigs, rotary rigs require drilling fluid to operate, and particles in the drilling fluid prevent formations from caving. The Lucas well showed that rotary rigs could drill wells that cable-tool rigs could not. Consequently, oilwell drillers began using rotary rigs more than cable-tool rigs. Today, almost all wells are drilled with rotary rigs. Because rotary rigs are so dominant, and because cable-tool rigs drilled a lot of wells before being supplanted by rotaries, let's find out more about them.

Cable-Tool and Rotary Drilling

3

Not counting picks and shovels, two drilling techniques have been available since people first began making holes in the ground: cable-tool drilling and rotary drilling. Both methods originated a long time ago. Over 2,000 years ago, for instance, the Chinese drilled wells with primitive yet efficient cable-tool rigs. (They were still using similar rigs as late as the 1940s.) To quarry rocks for the pyramids, the ancient Egyptians drilled holes using hand-powered rotating bits. They drilled several holes in a line and stuck dry wooden pegs in the holes. They then saturated the pegs with water. The swelling wood split the stone along the line made by the holes.

CABLE-TOOL DRILLING

Colonel Drake and Uncle Billy used a steam-powered cable-tool rig to drill the Oil Creek site. Early drillers in California and other parts of the world also used cable-tool rigs. To understand the principle of cable-tool drilling, picture a child's seesaw. Put a child on each end of it and let them rock it up and down. This rocking motion demonstrates the principle of cable-tool drilling. To explore it further, take the kids off the seesaw and go to one end of it. Tie a cable to the end and let the cable dangle straight down to the ground. Next, attach a heavy chisel with a sharp point to the dangling end of the cable. Adjust the cable's length so that when you hold the end of the seesaw all the way up, the chisel point hangs a short distance above the ground. Finally, let go of the seesaw. Releasing the seesaw lets the heavy chisel hit hard enough to punch a hole in the ground. Pick up the seesaw and repeat the process. Repeated rocking of the seesaw makes the chisel drill a hole. The process is quite effective. A heavy, sharp-pointed chisel can force its way through a great deal of rock with every blow.

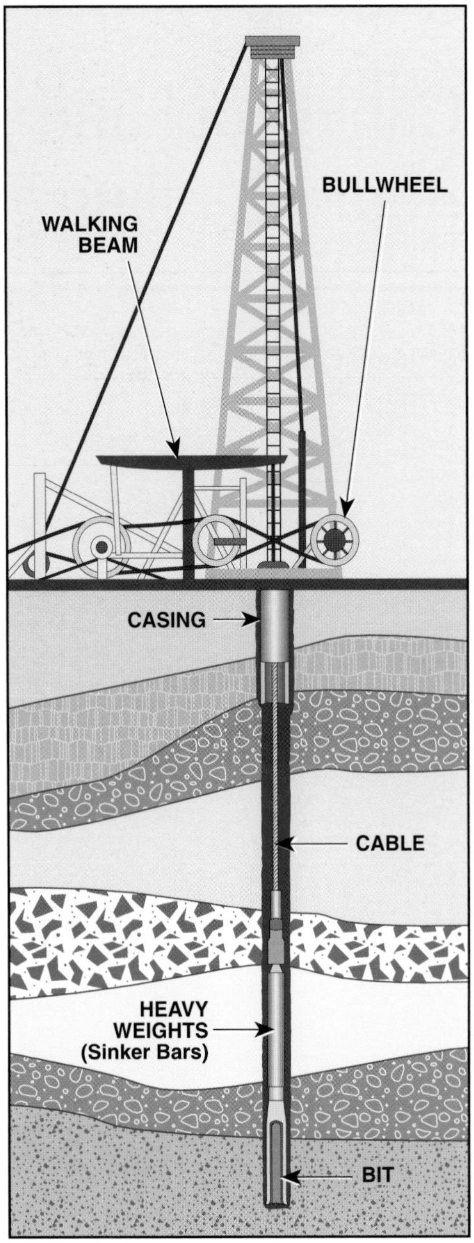

Figure 11. A cable-tool rig

A cable-tool rig (fig. 11) worked much like a seesaw. Of course, cable-tool rigs had more parts and, instead of a seesaw, a cable tool had a powered *walking beam* mounted in a derrick. At Drake's rig, a 6-horsepower (4.5-watt) steamboat engine powered the walking beam (fig. 12). The walking beam was a wooden bar that rocked up and down on a central pivot, much like a seesaw. The derrick provided a space to raise the cable and pull the long drilling tools out of the hole. As the beam rocked up it raised the cable and attached chisel, or bit. Then, when the walking beam rocked down, heavy weights, *sinker bars*, above the bit provided weight to ram it into the ground. The bit punched its way into the rock. Repeated lifting and dropping made the bit drill. Special equipment played out the cable as the hole deepened.

Figure 12. A 6-horsepower (4.5-watt) steamboat engine thought to be similar to Drake's original, powers the Drake well replica.

Cable-tool drilling worked very well in the hard-rock formations such as those in eastern U.S., the Midwest, and California. Indeed, a few cable-tool rigs are probably drilling wells somewhere in the world even now, although their use peaked in the 1920s and faded thereafter. Figure 13 pictures a 1920's cable-tool rig that drilled wells in Ohio and Pennsylvania until the 1950s.

In spite of cable-tool drilling's widespread use in the early days, the system had a couple of drawbacks. One was that cable-tool drillers had to periodically stop drilling and pull the bit from the hole. They then had to run a special basket, a *bailer*, into the hole to retrieve and remove the pieces of rock, or cuttings, the bit made. After bailing the cuttings, they then ran the bit back to bottom to resume drilling. If the crew failed to bail out the cuttings, the cuttings obstructed the bit's progress. Bailing cuttings was not a big hindrance, however, because the cable-tool system allowed the crew to do it quickly. Since the cable was wound onto a winch, or windlass, called the "bullwheel" (see fig. 11), the crew simply reeled cable on and off the bullwheel to raise and lower the bit and bailer. Reeling cable was a fast operation.

A far bigger problem than bailing, and the one that led to cable-tool drilling's demise, was that the cable-tool technique didn't work in soft formations like clay or loose sand. Clay and sand closed around the bit and wedged it in the hole. This limitation led to the increased use of rotary rigs because more wells were being drilled in places like Spindletop where cable-tool bits got stuck. The wall cake created by circulating drilling fluid prevented formations from collapsing.

Figure 13. A 1920's California standard cable-tool rig; it is located on the grounds of Drake Well State Park in northwestern Pennsylvania.

ROTARY DRILLING

Rotary drilling is quite different from cable-tool drilling. For one thing, a rotary rig uses a bit that isn't anything like a cable-tool's chisel bit. Instead of a chisel, a rotary bit has rows of teeth or other types of cutting devices that penetrate the formation and then scrape or gouge out pieces of it as the rig system rotates the bit (fig. 14). Further, a rotary rig doesn't use cable to suspend the bit in the hole. Rotary crew members attach the bit to the end of a long string of hollow pipe. By screwing together several joints of pipe, they put the bit on the bottom of the hole (fig. 15). As the hole deepens, they add joints of pipe (fig. 16).

Figure 14. A bit for a rotary drilling rig

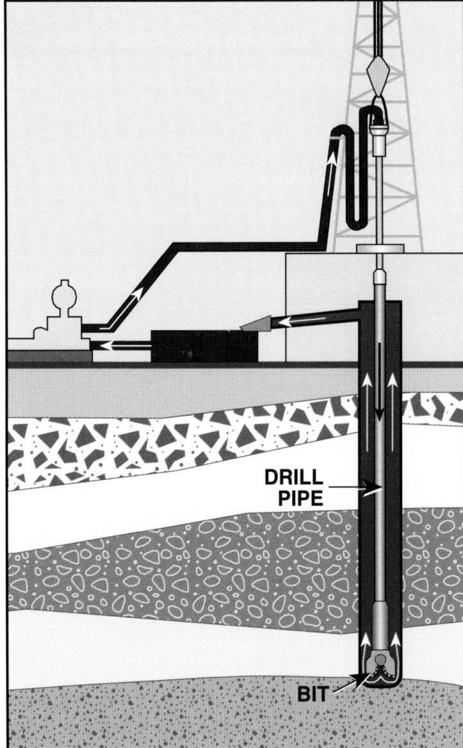

Figure 15. The drill stem puts the bit on bottom to drill.

Rotating Systems

With the bit on bottom, the rig can rotate it in one of three ways. Many rigs use a machine called a "rotary table," a sort of heavy-duty turntable (fig. 17). Others rotate the bit with a *top drive*, a device with a powerful built-in electric motor that turns the pipe and bit (fig. 18). And, in special cases, a slim *downhole motor*,

Figure 16. Two rotary helpers lift a joint of drill pipe out of the mousehole prior to adding it to the active drill string.

Figure 17. Components in the rotary table rotate the drill string and bit.

Figure 18. A powerful motor in the top drive rotates the drill string and bit.

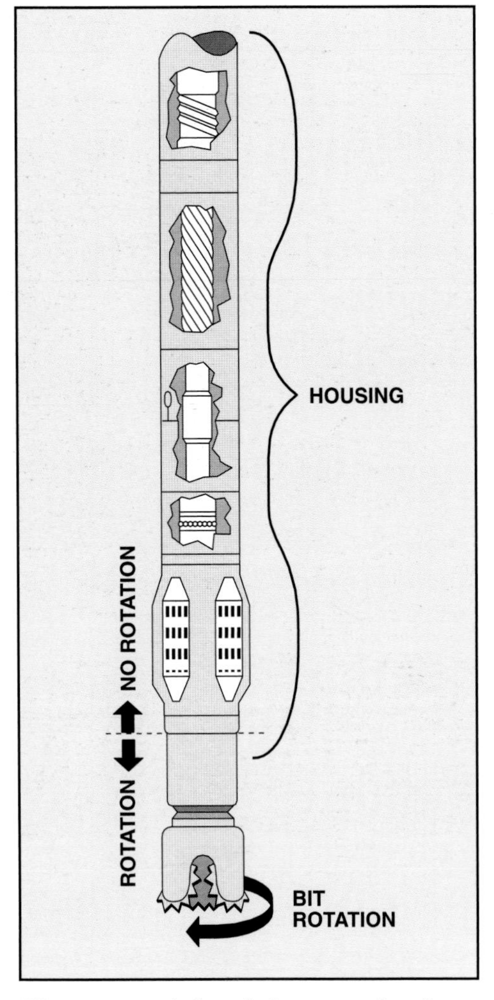

Figure 19. A downhole motor placed near the bit rotates the bit.

usually powered by drilling fluid but in some cases by electricity, rotates the bit (fig. 19). A long metal housing with a diameter a little less than the hole's holds the motor. The bit screws onto the end of it.

Generally, the latest rotary rigs use a top drive to rotate the pipe and bit. However, rigs using rotary tables have been around a long time and many drilling companies still own and use them. Moreover, rotary tables are simple, rugged, and easy to maintain. Rotary rig owners often use downhole motors where they have to rotate the bit without rotating the entire string of pipe. Such situations occur when the rig is drilling a slant, or *directional hole*, a hole that is intentionally diverted from vertical to better exploit a reservoir. (A later chapter in this book covers directional drilling in more detail.)

Regardless of the system used to rotate the bit, the *driller*, the person operating the rig, allows some of the weight of the pipe to press down on the bit. The weight causes the bit's cutters to bite into the formation rock. Then, as the bit rotates, the cutters roll over the rock and scrape or gouge it out.

Fluid Circulation

By itself, rotating a bit on pipe does not get the job done. The cuttings the bit makes must be moved out of the way. Otherwise, they collect under the bit cutters and impede drilling. Recall that the crew on a cable-tool rig had to stop drilling and bail the cuttings. A rotary rig crew does not have to bail cuttings, because the rig circulates fluid while the bit drills and the fluid carries the cuttings up to the surface. As mentioned earlier, crew members attach a rotary bit to hollow pipe, instead of to braided cable. The pipe is thus a conduit: a powerful pump on the surface (fig. 20) moves fluid down the pipe to the bit and back to the surface (fig. 21). This fluid picks up the cuttings as the bit makes them and carries them to the surface where they are disposed of. The pump then moves the clean mud back down the hole.

Figure 20. Two pumps are available on this rig to move drilling fluid down the pipe; normally only one at a time is used. If more volume is needed, however, the other pump can also be put into service. (Some very large rigs have three or four pumps.)

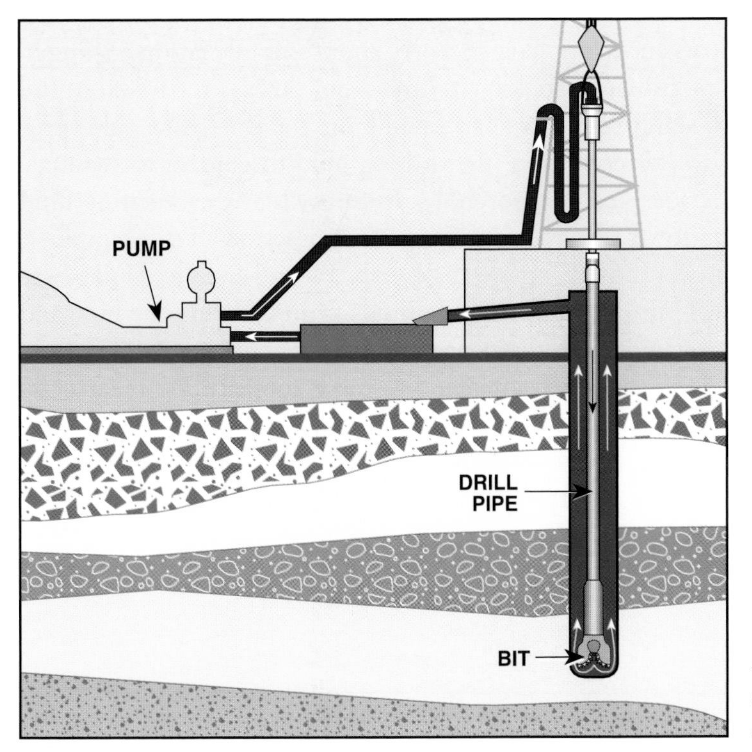

Figure 21. A pump circulates drilling mud down the drill pipe, out the bit, and up the hole.

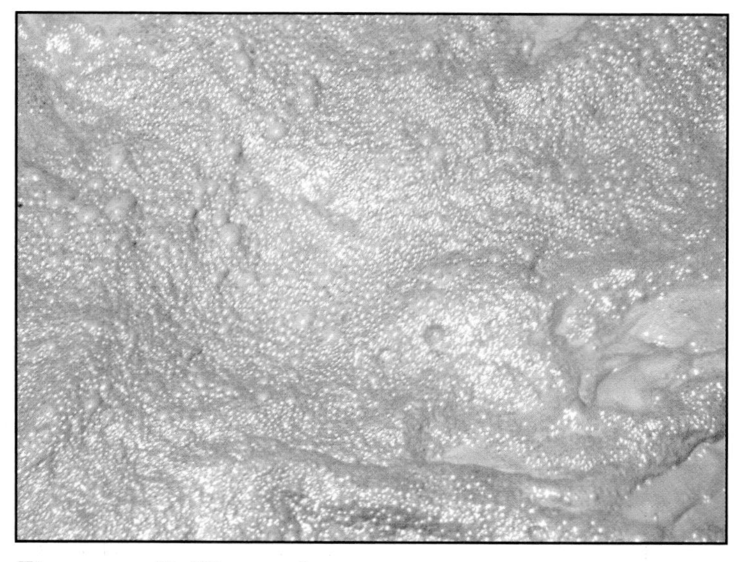

Figure 22. Drilling mud

The fluid is usually a special liquid called "drilling mud" (fig. 22). Don't be misled by the name, however. Although the earliest drilling muds were not much more than a plain, watery mud (recall that the Hamil brothers supposedly filled a pit with water and ran cattle through it to make it muddy), drilling mud can be a complex blend of materials. What's more, sometimes it isn't a liquid, which is why a better name for drilling mud is "drilling fluid." A fluid can be a liquid, a gas, or a combination of the two.

As you now know, one advantage of a rotary rig is that workers do not have to worry about soft formations caving in on the bit and sticking it. Just as the Hamils prepared the mud to stabilize the hole at Spindletop, today's drillers also prepare, or condition, the drilling mud to control formations. Besides keeping boreholes from caving in, circulating mud performs several other important functions. For example, it moves the cuttings away from the bit and cools and lubricates it. It also keeps formation fluids from entering the hole and blowing out to the surface. Indeed, circulating drilling fluid has so many advantages that cable-tool drilling is virtually obsolete. Although companies may use a cable-tool rig in a few special cases, more often they use rotary rigs. Several kinds of rotary rig are available for drilling on land and offshore. Let's look at the major types.

4

Rotary Rig Types

Many kinds of rotary drilling rigs are available, particularly offshore where the marine environment plays an important role in rig design. Two broad categories of rig are those that work on land (fig. 23) and those that work offshore (fig. 24). Some experts like to create a third category: rigs that work in inland waters (fig. 25). Inland rigs usually drill in lakes, marshes, and estuaries, places that are neither land nor offshore, places where, as one wit put it, "it's too wet to plow and too muddy to drink." For our purposes, though, dividing rotary rigs into land and offshore types is acceptable, because inland rigs also drill in water, even if it is shallow.

Figure 23. A land rig

Figure 24. An offshore rig

Figure 25. An inland barge rig

Table 1
Land rigs classified by drilling depth

Rig Size	Maximum Drilling Depth, Feet (Metres)
Light duty	3,000–5,000 (1,000–1,500)
Medium duty	4,000–10,000 (1,200–3,000)
Heavy duty	12,000–16,000 (3,500–5,000)
Very heavy duty	18,000–25,000+ (5,500–7,500+)

LAND RIGS

Land rigs look much alike, although details vary. A major difference is their size, and size determines how deep the rig can drill. Well depths range from a few hundred or thousand feet (metres) to tens of thousands of feet (metres). The depth of the formation that contains, or is believed to contain, oil and gas controls well depth. Classified by size, land rigs are light duty, medium duty, heavy duty, and very heavy duty. Table 1 arranges them according to this scheme and shows the depths to which they can drill.

Keep in mind, though, that a rig can drill holes shallower than its maximum rated depth. For example, a medium-duty rig could drill a 2,500-foot (750-metre) hole, although a light-duty rig could also drill it. On the other hand, a rig cannot drill too much beyond its rated maximum depth, because it cannot handle the heavier weight of the drilling equipment required for deeper holes.

Another feature land rigs share is portability. A rig can drill a hole at one site, be disassembled if required, moved to another site (fig. 26), and be reassembled to drill another hole. Indeed, land rigs are so mobile that one definition terms them "portable hole factories." The definition sounds odd, but it is accurate.

Figure 26. To move a rig, crew members disassemble it and move it piece by piece.

MOBILE OFFSHORE RIGS

A widely used offshore drilling rig is a *mobile offshore drilling unit*, or *MODU*, for short (pronounced "mow-du"). Another is a *platform*. Although drilling occurs from platforms, companies mainly employ them on the producing side of the oil and gas business. This book concentrates on drilling, so it does not cover platforms. However, more information about platforms is available in the PETEX publication, *A Primer of Offshore Operations.*

MODUs are portable; they drill a well at one offshore site and then move to drill another. MODUs are either *floaters* or *bottom-supported.* When drilling, floaters work on top of, or slightly below, the water's surface. Floaters include *semisubmersibles* and *drill ships*. They are capable of drilling in waters thousands of feet (metres) deep. MODUs that contact the ocean bottom and are supported by it are bottom-supported. Bottom-supported units include *submersibles* and *jackups.* Submersibles are further divided into *posted barges, bottle types, inland barges*, and *arctic.* Generally, bottom-supported rigs drill in waters shallower than floaters. Table 2 lists MODUs.

Table 2
Types of MODU

Mobile Offshore Drilling Units (MODUs)	
Bottom-Supported Units	*Floating Units*
Submersibles	Semisubmersibles
Posted barges	Drill ships
Bottle types	
Arctic types	
Inland barges	
Jackups	

Bottom-Supported Units

Submersibles and jackups contact the seafloor when drilling. The lower part of a submersible's structure rests on the seafloor. In the case of jackups, only the legs contact the seafloor.

Submersibles

A submersible MODU floats on the water's surface when moved from one drilling site to another. When it reaches the site, crew members flood compartments that submerge the lower part of the rig to the seafloor. With the base of the rig in contact with the ocean bottom, wind, waves, and currents have little effect on it.

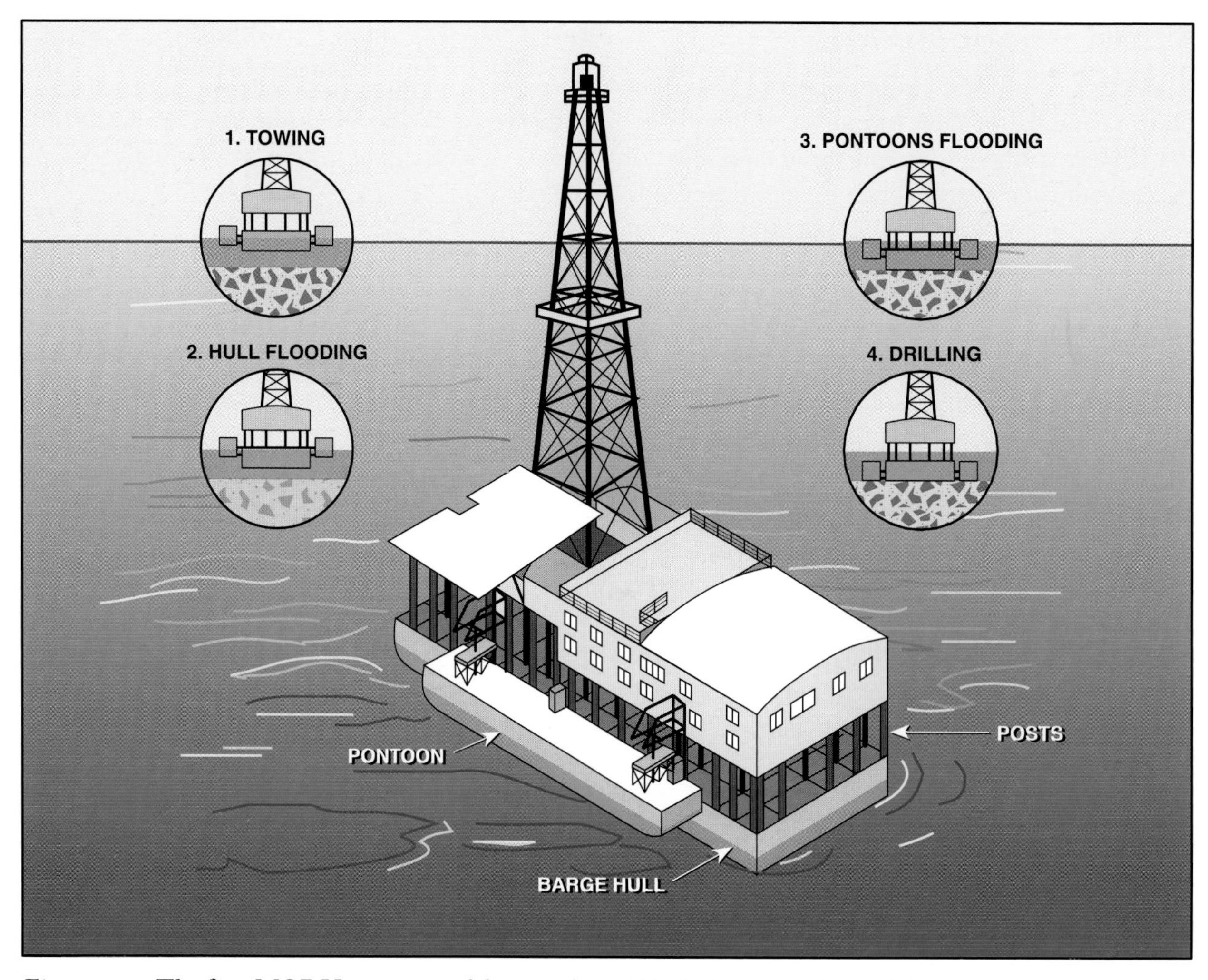

Figure 27. The first MODU was a posted-barge submersible designed to drill in shallow waters.

Posted-Barge Submersibles

The first MODU was a submersible. It drilled its initial well in 1949 off the Gulf Coast of Louisiana in 18 feet (5.5 metres) of water. It was a posted-barge submersible—a barge hull and steel posts (columns) supported a deck and drilling equipment (fig. 27). It proved that mobile rigs could drill offshore. Posted barges are now virtually obsolete, however, because newer and better designs have replaced them.

Bottle-Type Submersibles

About 1954, drilling moved into water depths beyond the posted barge's capabilities, which was about 30 feet (9 metres).

So, naval architects designed bottle-type submersibles. A bottle-type rig has four tall steel cylinders (bottles) at each corner of the structure. The main deck lies across several steel supports and the bottles. The rig and other equipment are placed on the main deck. When flooded, the bottles cause the rig to submerge to the seafloor (fig. 28).

In their heyday in the early 1960s, the biggest bottle-type submersibles drilled in 150-foot (45-metre) water depths. Today, jackups have largely replaced them; jackups are less expensive to build than bottle-types and can drill in deeper water. Rather than completely scrap their bottle types, however, rig owners modified some of them to drill as semisubmersibles, which are still in use. (Semisubmersibles are covered shortly.)

Figure 28. When flooded, the bottles cause a bottle-type submersible to submerge to the seafloor.

Arctic Submersibles

A special type of submersible rig is an arctic submersible. In the arctic, where petroleum deposits lie under shallow oceans such as the Beaufort Sea, oil companies knew that jackups and conventional barge rigs would not be suitable. During the arctic winter, massive chunks of ice form and then move with currents on the water's surface. Called "floes," these moving ice blocks exert tremendous force on any object they contact. The force is great enough to destroy the legs of a jackup or the hull of a conventional ship or a barge.

Arctic submersibles therefore have a reinforced hull, a *caisson*. One type of caisson has a reinforced concrete base on which the drilling rig is installed (fig. 29). When the sea is

Figure 29. A concrete island drilling system (CIDS) features a reinforced concrete caisson.

ice-free in the brief arctic summer, boats tow the submersible to the drilling site. There, workers submerge the caisson to the sea bottom and start drilling. Shortly, when ice floes form and begin to move, the arctic submersible's strong caisson hull deflects the floes, enabling operations to continue.

Inland Barge Rigs

A fourth submersible is an inland barge rig. It has a barge hull—a flat-bottomed, flat-sided, rectangular steel box. The rig builder places a drilling rig and other equipment on the barge deck (fig. 30). Inland barge rigs normally drill in marshes, bays, swamps, or other shallow inland waters. By definition, barges are not *self-propelled*; they have no built-in power to move them from one site to another. Therefore, boats tow them to the drilling location. When being moved, the barge floats on the water's surface; then, when positioned at the drilling site, the barge is flooded so that it rests on the bottom ooze. Since they often drill in swampy shallow waters, drilling people often call inland barges "swamp barges."

Figure 30. Drilling equipment is placed on the deck of a barge to drill in the shallow waters of bays and estuaries.

Jackups

A jackup rig is a widely used mobile offshore drilling unit. It floats on a barge hull when towed to the drilling location (fig. 31). Most modern jackups have three legs with a triangular-shaped barge hull; others have four or more legs with rectangular hulls. A jackup's legs can be cylindrical columns, somewhat like pillars (fig. 32), or they can be *open-truss structures*, which resemble a mast or a derrick (fig. 33).

Figure 31. Four boats tow this jackup to its drilling location.

Figure 32. A jackup rig with four columnar legs

Figure 33. This jackup has open-truss legs.

Figure 34. The hull of this jackup is raised to clear the highest anticipated waves.

Whether it has columnar or open-truss legs, when a jackup's barge hull is positioned on the drilling site, the crew jacks down the legs until they contact the seafloor. They then raise, or jack up, the hull above the height of the highest anticipated waves (fig. 34). The drilling equipment is on top of the hull. The largest jackups can drill in water depths up to about 400 feet (about 120 metres), and are capable of drilling holes up to 30,000 feet (10,000 metres), or close to 5½ miles, deep.

Floating Units

Floating offshore drilling rigs include semisubmersibles and drill ships. Semisubmersibles, because of their design, are more stable than drill ships. On the other hand, drill ships can carry more drilling equipment and supplies, which often make them the choice in remote waters.

Semisubmersibles

Most semisubmersible rigs have two or more *pontoons* on which the rig floats. A pontoon is a long, relatively narrow, and hollow steel float with a rectangular or round cross section (fig. 35). When a semisubmersible is moved, the pontoons contain mostly air so that the rig floats on the water's surface. In most cases, towboats then tie onto the rig and move it to the drill site. However, some semisubmersible rigs are self-propelled—they have built-in power units that drive the rig from one site to another.

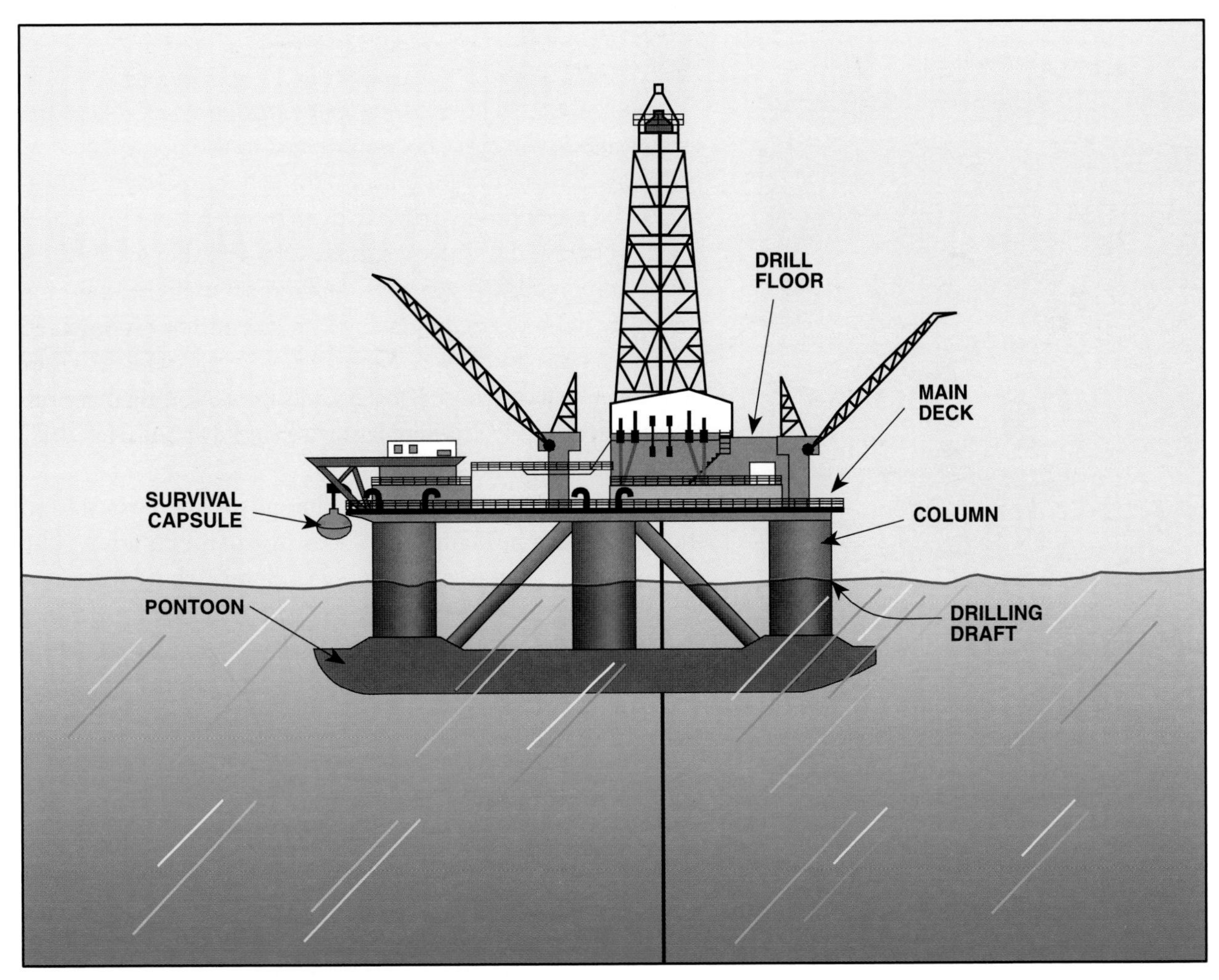

Figure 35. A semisubmersible rig floats on pontoons.

Figure 36. The pontoons of this semisubmersible float a few feet (metres) below the water's surface.

Figure 37. The main deck of a semisubmersible is big.

Semisubmersibles get their name from the fact that in the drilling mode the rig is not submerged to the point where its pontoons contact the sea bottom. Instead, rig personnel carefully flood the pontoons to make them submerge only a few feet (metres) below the water's surface (fig. 36). Thus, the rig is "semisubmerged." (If the pontoons contacted the sea bottom, the rig would be "submerged.") With its pontoons submerged below the waterline, waves do not affect the rig as much as they do when it floats on the surface. A semisubmersible rig therefore offers a more stable drilling platform than a drill ship that drills while floating on the water's surface.

Large cylindrical or square columns extend upward from the pontoons. The main deck rests on top of the columns. The main deck of a semi is big (fig. 37). Semis (short for semisubmersibles) often use anchors to keep them on the drilling station. Workers release several large anchors from the deck of the rig. An anchor-handling boat crew sets the anchors on the seafloor.

Besides being good rough-water rigs, semis are also capable of drilling in water thousands of feet (metres) deep. While many semis work in water depths ranging from 1,000 to 3,500 feet (300 to 1,000 metres), the latest are capable of drilling in water depths of 8,000 feet (2,500 metres). Semis can drill holes up to 30,000 feet (10,000 metres) deep. Indeed, semisubmersibles are among the largest floating structures ever made.

The biggest ones soar to over 100 feet (30 metres) tall and their main decks can be almost as big as a football field—that's 3,000 square yards (2,500 square metres).

Drill Ships

A drill ship is also a floater (fig. 38). Drill ships are very mobile because they are self-propelled and have a streamlined hull, much like a regular ocean-going ship. A company may therefore choose a drill ship to make hole in remote waters, far from land. A drill ship is a good choice for drilling remote locations. For one thing, it can move at reasonable speeds under its own power. Secondly, its ship-shaped hull can carry a large amount of the equipment and material required for drilling. Frequent resupplying from a shore base is therefore not necessary.

While many drill ships operate in water depths ranging from 1,000 to 3,000 feet (300 to 1,000 metres), the latest can drill in water depths approaching 10,000 feet (3,000 metres), or nearly 2 miles (3.2 kilometres). They can drill holes over 30,000 feet (10,000 metres) deep. These big drill ships are more than 800 feet (250 metres) long, which is almost as long as three football fields laid end to end. They measure some 100 feet (30 metres) wide, or a little wider than a football field.

Figure 38. A drill ship

Their hulls tower more than 60 feet (18 metres) high, about that of a six-story building.

Anchors keep some drill ships on station while drilling, but those drilling in deep water require *dynamic positioning*. Dynamically positioned drill ships use computer-controlled *thrusters* and sophisticated electronic sensors. Thrusters are power units with propellers that the builder mounts fore and aft on the drill ship's hull below the waterline. Once the dynamic positioning operator tells a computer exactly where it should keep the rig positioned, the computer, using information transmitted by the sensors, automatically controls the thrusters. The thrusters offset wind, wave, and current forces that would move the rig away from the desired position.

Whether on land or offshore, and whether large, medium, or small, all rigs require personnel to operate them. The people who drill wells usually work for a company whose business involves drilling, either directly or indirectly. So, let's look next at companies involved in drilling and the personnel who do the work.

5 People and Companies

People who work for companies involved in drilling work all over the world. They drill wells on land and ice, in swamps, and on waters as small as lakes or as large as the Pacific Ocean. Drilling is demanding; it goes on 24 hours a day, 7 days a week, in all types of weather. Moreover, drilling is complex; so complex that no single company is diverse enough to perform all the required work. Consequently, many companies and individuals are involved. Companies include operating companies, drilling contractors, and service and supply companies.

OPERATING COMPANIES

An *operating company*, or an *operator*, is usually an oil company, a company whose primary business is working with oil and gas, or petroleum. An operating company may be an *independent* or a *major*. An independent company may be one or two individuals or it may have hundreds of employees. Major companies, such as ExxonMobil, BP Amoco, or Shell, may have thousands of employees. Besides size, another difference between an independent and a major is that, in general, an independent only produces and sells crude oil and natural gas. A major, on the other hand, produces crude oil and natural gas, transports them from the field to a refinery or a plant, refines or processes the oil and gas, and sells the products to consumers.

Whether independent or major, an operator must acquire the right to drill for and produce petroleum at a particular site. An operating company does not usually own the land or the minerals (oil and gas are minerals) lying under the land. It therefore has to buy or lease the rights to drill for and produce oil and gas from the landowner and the mineral holder. Individuals, partnerships, corporations, or a federal, state, or local government can own land and mineral rights. The operator not only pays the landowner a fee for leasing, it also pays the mineral holder a royalty, which is a share of the money made from the sale of oil or gas.

DRILLING CONTRACTORS

Drilling is a unique undertaking that requires experienced personnel and special equipment. Most operating companies therefore find it more cost effective to hire expertise and equipment from drilling companies than to keep the personnel and equipment under their own roof. So, almost everywhere in the world, *drilling contractors* do the drilling.

A drilling contractor is an individual or a company that owns from one to dozens of drilling rigs. The contractor hires out a rig and the personnel needed to run it to any operator who wishes to pay to have a well drilled. Some contractors are land contractors—they operate only land rigs. Others are offshore contractors—they operate only offshore rigs. A few contractors operate rigs that drill both on land and offshore. The contractor may have different sizes of rigs that can drill to various depths. A drilling contracting company may be small or large; it may own rigs that drill mainly in a local area or it may have rigs working all over the world.

Regardless of its size, a drilling company's job is to drill holes. It must drill holes to the depth and specifications set by the operating company, who is also the well owner. An operating company usually invites several contractors to bid on a job. Often, the operator awards the contract to the lowest bidder, but not always. Sometimes a good work record may override a low bid.

DRILLING CONTRACTS

The operator usually sends a proposal to several drilling contractors. The proposal describes the drilling project and requests a bid. The contractor then fills out the proposal, signs it, and sends it back to the operator. If the operating company accepts the bid, it becomes a contract between the operator and the drilling company. This signed agreement clearly states the services and supplies the contractor and the operator are to provide for a particular project.

The International Association of Drilling Contractors (IADC) supplies popular contract forms (fig. 39). IADC is an organization whose membership is made up of drilling contractors, oil companies, and service and supply companies with an interest in drilling. Headquartered in Houston, Texas, and with offices throughout the world, IADC provides many services to its members, not only in the U.S., but also in other parts of the globe. Its mission is "to promote a commitment to safety, to preservation of the environment, and to advances in drilling technology."

NOTE: This form contract is a suggested guide only and use of this form or any variation thereof shall be at the sole discretion and risk of the user parties. Users of the form contract or any portion or variation thereof are encouraged to seek the advice of counsel to ensure that their contract reflects the complete agreement of the parties and applicable law. The International Association of Drilling Contractors disclaims any liability whatsoever for loss or damages which may result from use of the form contract or portions or variations thereof.

Revised July, 1998

INTERNATIONAL ASSOCIATION OF DRILLING CONTRACTORS

DRILLING BID PROPOSAL AND DAYWORK DRILLING CONTRACT - U.S.

TO: ______________________
____

Please submit bid on this drilling contract form for performing the work outlined below, upon the terms and for the consideration set forth, with the understanding that if the bid is accepted by ______________________ this instrument will constitute a contract between us. Your bid should be mailed or delivered not later than ________ P.M. on ______________, 19_____ to the following address: ______________________

THIS AGREEMENT CONTAINS PROVISIONS RELATING TO INDEMNITY, RELEASE OF LIABILITY, AND ALLOCATION OF RISK

THIS AGREEMENT (The "Contract") is made and entered into on the date hereinafter set forth by and between the parties herein designated as "Operator" and "Contractor".

OPERATOR: ______________________
Address: ______________________

CONTRACTOR: ______________________
Address: ______________________

IN CONSIDERATION of the mutual promises, conditions and agreements herein contained and the specifications and special provisions set forth in Exhibit "A" and Exhibit "B" attached hereto and made a part hereof, Operator engages Contractor as an Independent Contractor to drill the hereinafter designated well or wells in search of oil or gas on a daywork basis.

For purposes hereof, the term "daywork basis" means Contractor shall furnish equipment, labor, and perform services as herein provided, for a specified sum per day under the direction, supervision and control of Operator (inclusive of any employee, agent, consultant or subcontractor engaged by Operator to direct drilling operations). When operating on a daywork basis, Contractor shall be fully paid at the applicable rates of payment and assumes only the obligations and liabilities stated herein. Except for such obligations and liabilities specifically assumed by Contractor, Operator shall be solely responsible and assumes liability for all consequences of operations by both parties while on a daywork basis, including results and all other risks or liabilities incurred in or incident to such operations.

1. **LOCATION OF WELL:**
Well Name and Number: ______________________
Parish/County: ______________ State: ______________ Field Name: ______________________
Well location and land description: ______________________

1.1 **Additional Well Locations or Areas:** ______________________

Locations described above are for well and Contract identification only and Contractor assumes no liability whatsoever for a proper survey or location stake on Operator's lease.

2. **COMMENCEMENT DATE:**
Contractor agrees to use reasonable efforts to commence operations for the drilling of the well by the ________ day of ______________, 19_____, or ______________________

3. **DEPTH:**
3.1 **Well Depth:** The well(s) shall be drilled to a depth of approximately ______________ feet, or to the ______________ formation, whichever is deeper, but the Contractor shall not be required hereunder to drill said well(s) below a maximum depth of ______________ feet, unless Contractor and Operator mutually agree to drill to a greater depth.

4. **DAYWORK RATES:**
Contractor shall be paid at the following rates for the work performed hereunder.
4.1 **Mobilization:** Operator shall pay Contractor a mobilization fee of $______________ or a mobilization day rate of $______________ per day. This sum shall be due and payable in full at the time the rig is rigged up or positioned at the well site ready to spud. Mobilization shall include: ______________________

4.2 **Demobilization:** Operator shall pay Contractor a demobilization fee of $______________ or a demobilization day rate during tear down of $______________ per day, provided however that no demobilization fee shall be payable if the Contract is terminated due to the total loss or destruction of the rig. Demobilization shall include: ______________________

4.3 **Moving Rate:** During the time the rig is in transit to or from a drill site, or between drill sites, commencing on ______________, Operator shall pay Contractor a sum of $______________ per twenty-four (24) hour day.

4.4 **Operating Day Rate:** For work performed per twenty-four (24) hour day with ______________ man crew the operating day rate shall be:

Depth Intervals			
From	To	Without Drill Pipe	With Drill Pipe
__________	__________	$__________ per day	$__________ per day
__________	__________	$__________ per day	$__________ per day
__________	__________	$__________ per day	$__________ per day

Using Operator's drill pipe $______________ per day.

(U.S. Daywork Contract - Page 1)

Figure 39. An IADC drilling bid form; when signed by all parties, it becomes a contract.

Contractors are paid for the work their rig and crews do in several ways. Operators can pay contractors based on the daily costs of operating the rig, the number of feet or metres drilled, or on a turnkey basis. If the contractor is paid according to the daily costs of operating the rig, it's a *daywork contract*. If the contract calls for the contractor to be paid by the number of feet or metres drilled, it's a *footage* or *metreage contract*. And, as you can guess, if it's a turnkey job, then the operator and contractor sign a *turnkey contract*, in which the drilling contractor is responsible for the entire drilling operation. Daywork contracts are the most common.

SERVICE AND SUPPLY COMPANIES

The operating company owns the well and usually hires a drilling contractor to drill it. But to successfully drill a well, the operator and the contractor need equipment, supplies, and services that neither company normally keeps on hand. So, service and supply companies provide the required tools and services to expedite the drilling of the well. *Supply companies* sell expendable and nonexpendable equipment and material to the operator and the drilling contractor. Expendable items include drill bits, fuel, lubricants, and drilling mud—items that are used up or worn out as the well is drilled. Nonexpendable items include drill pipe, fire extinguishers, and equipment that may eventually wear out and have to be replaced but normally last a long time. Likewise, supply companies market safety equipment, rig components, tools, computers, paint, grease, rags, and solvents. Think of any part or commodity that a rig needs to drill a well, and you'll find a supply company on hand to provide it.

Service companies offer special support to the drilling operation. For example, a *mud logging company* monitors and records, or *logs*, the content of the drilling mud as it returns from the well. The returning mud carries cuttings and any formation fluids, such as gas or oil, to the surface. The operator can gain knowledge about the formations being drilled by analyzing the returning drilling fluid.

In many instances, when a well reaches a formation of interest (usually, a formation that may contain oil or gas), the operator hires a *well logging company*. A logging crew runs sophisticated instruments into the hole. These instruments sense and record formation properties. Computers in the field generate special graphs, called "well logs," for the operator to examine (fig. 40). Well logs help the operating company determine whether the well will produce oil or gas.

Figure 40. This computer display shows a well log.

Figure 41. This member of a casing crew is stabbing one joint of casing into another. The red fitting is a casing coupling used to connect the joints.

Another service company provides *casing crews*. A casing crew runs special pipe, casing, into the well to line, or *case*, it (fig. 41) after the rig drills a portion of the hole. Casing protects formations from contamination and stabilizes the well. After the casing crew runs the casing, another service company—a *cementing company*—cements the casing in the well. Cement bonds the casing to the hole.

Most offshore rigs, and land rigs in very remote fields, require cooking and housekeeping services, since personnel live as well as work offshore or in isolated regions for long periods (fig. 42). The drilling contractor or operating company often hires an oilfield caterer to furnish these services.

Figure 42. Personnel on this offshore rig enjoy good food in the galley; a catering company usually provides the food and cooking.

PEOPLE

While it is true that you can't drill a well without a drilling rig and several companies to back up the rig, it is equally true that you can't drill a well without skilled people. Personnel run the rig and keep it running until the well reaches its objective. Many people are involved in drilling. Let's cover the drilling crew first—the group whose job it is to make the rig drill.

Drilling Crews

The contractor requires trained and skilled personnel to operate and maintain a drilling rig. Keep in mind that a rig, when on site and drilling, operates virtually all the time, night and day, 365 days a year. Personnel directly responsible for making the rig drill are collectively known as the "drilling crew."

The person in charge of the drilling crew, the top hand, may be called the "rig manager," "rig superintendent," or "toolpusher," depending on the drilling contractor's preference. Besides the rig manager, or superintendent, each rig has *drillers*, *derrickmen*, and *rotary helpers* (also called "floorhands," or "roughnecks"). What's more, large land rigs and offshore rigs often have *assistant rig supervisors*, *assistant drillers*, as well as additional personnel who perform special functions particular to the rig.

Rig Superintendent and Assistant Rig Superintendent

The rig superintendent (rig manager or toolpusher) oversees the drilling crews that work on the rig floor, supervises drilling operations, and coordinates operating company and contractor affairs. On land rigs, the rig superintendent is usually headquartered in a mobile home or a portable building at the rig site and is on call at all times. Offshore, the rig superintendent has an office and sleeping quarters on the rig, and is also on call at all times. Because offshore drilling and large land drilling operations can be very critical, the contractor may hire an assistant rig superintendent. The assistant rig superintendent often relieves the superintendent during nighttime hours and is thus sometimes nicknamed the "night toolpusher."

Figure 43. The driller on this offshore rig works in an environmentally controlled cabin.

Driller and Assistant Driller

The rig superintendent supervises the driller, who, in turn, supervises the derrickman and the rotary helpers. From a control console or an operating cabin on the rig floor, the driller manipulates the controls that keep the drilling operation under way (fig. 43). This person is directly responsible for drilling the hole. Most offshore rigs and large land rigs, especially those working outside the U.S., also have an *assistant driller*. The assistant driller aids the driller on the rig floor and helps the driller supervise the derrickman and the rotary helpers.

Derrickman

A few of the latest rigs feature automated pipe-handling equipment that takes over the duties of the *derrickman*. Most rigs, however, require a derrickman when crew members run drill pipe into the hole (when they *trip in*), or when they pull pipe out of the hole (when they *trip out*). The derrickman handles the upper end of the pipe from the *monkeyboard* (fig. 44). The monkeyboard is a small platform in the mast or derrick on which the derrickman stands to handle the upper end of the pipe. The contractor mounts the monkeyboard in the mast or derrick at a height ranging from about 50 to 110 feet (15 to 34 metres), depending on the length of the joints of pipe crew members pull from the hole. The derrickman uses special safety equipment to prevent falls.

Figure 44. This picture was shot above the derrickman's position on the monkeyboard. Drill pipe is racked to the left. Drilling line runs from top to the traveling block below. Note that this rig has eight lines strung.

In addition, the derrickman has an escape device, a *Geronimo*, or a *Tinkerbell line*, so that he or she can quickly exit the monkeyboard in an emergency. (Geronimo was a Chiricahua Apache who eluded the Army for many years in the American southwest in the late 1800s. For some reason, World War II paratroopers sometimes yelled his name when they jumped out of airplanes. Tinkerbell is a fictitious flying character from the children's novel *Peter Pan*.) In any case, if the derrickman has to get out of the derrick or mast quickly, he or she grasps a handle on the Geronimo and rides it down on a special cable, or line, to the ground. The derrickman controls the rate of descent by moving the handle to increase or decrease braking action on the line.

When the bit is drilling and the pipe is in the hole, the derrickman, using a built-in ladder in the derrick or mast for normal descent, climbs down from the monkeyboard and works at ground level. When not in the derrick or mast, derrickmen monitor the condition of the drilling mud (fig. 45). They make sure it meets the specifications for drilling a particular part of the hole.

Figure 45. This derrickman is checking the weight, or density, of the drilling mud.

Rotary Helpers (Floorhands)

Depending on the size of the rig, its equipment, and other factors, a contractor usually hires two or three *rotary helpers*, or *floorhands*, for each work shift. On small rigs drilling shallow wells, for example, two rotary helpers on a shift can safely and efficiently perform the required duties. On large

Figure 46. Two rotary helpers latch big wrenches (tongs) onto the drill pipe.

rigs drilling deep holes, and offshore, the job usually requires three rotary helpers, but not always. In either case, on conventional rigs, rotary helpers handle the lower end of the drill pipe when they are tripping it in or out of the hole. They also use large wrenches called "tongs" to screw or unscrew (*make up* or *break out*) pipe (fig. 46). Some tongs are *power tongs* (fig. 47), which replace conventional tongs. Besides handling pipe, rotary helpers also maintain the drilling equipment, help repair it, and keep it clean and painted.

Rotary helpers get their name from the fact that much of their work occurs on the rig floor, near the rotary table—the traditional device that turns the drill pipe and bit. Originally, they were also called "roughnecks," probably because those who worked on early rigs prided themselves in being rough and tough. Later, they became rotary helpers, which added a little dignity to the title. They are also called floorhands because they perform most of their duties on the rig floor.

Figure 47. These two floorhands are using power tongs to tighten drill pipe.

Drilling Crew Work Shifts

Because of a rig's location, economic factors, and other reasons, the number of days and the number of hours per day that a drilling crew works vary a great deal.

Regardless of the length of their workday, drilling crews call their shifts "tours." Strangely, they pronounce tour as "tower." This odd pronunciation is traditional and apparently began when a not-too-well-read rig hand saw the word "tour," as in tour of duty, and mispronounced it with two syllables. However, "too-ur" (or whatever the pronunciation was) must have been difficult to say, for rig crews everywhere began pronouncing it "tower," and it stuck.

In a few areas, particularly in West Texas and Eastern New Mexico, contractors employ 8-hour tours. In other areas, such as offshore, along the Gulf Coast, in countries outside the U.S., and in remote land locations, they use 12-hour tours. If the crews work 8-hour tours, then the contractor usually hires four drilling crews and two toolpushers, or rig superintendents, for each rig. The crews consist of four drilling crews—four drillers and derrickmen, and eight or 12 rotary helpers. Three drilling crews split three 8-hour tours per day. The fourth crew is off. Later, they relieve one of the working crews. One rig superintendent, or toolpusher, is on the site all the time. He or she may work 7 days, for example, and then be relieved by the other superintendent for 7 days.

If the crews work 12-hour tours on land, then the contractor may hire two drilling crews and two superintendents for each rig. One superintendent, two drillers, two assistant drillers (if the rig requires them), two derrickmen, and four or six rotary helpers—two full drilling crews—split two tours per 24-hour day.

Offshore, crews also usually work 12-hour tours, but the contractor hires four drilling crews. Two crews may work 14 days and then take off 14 days when the second crews come on board to relieve them. Some contractors based in the U.S. have rigs working abroad, such as in the North Sea or in Southeast Asia. In such cases, the contractor often employs a 28-and-28 schedule. Two crews are home for 28 days while the other two work 12-hour tours for 28 days.

Other Rig Workers

Besides the drilling crew, many other persons work at the rig site. They may be there during the entire time the well is being drilled, or they may come out only when their expertise or equipment is needed.

The Company Representative

The operating company customarily has an employee on the drill site to supervise its interests. The *company representative*, or *company man*, on a land rig, like the rig superintendent, usually lives on the rig site in a mobile home or portable building. Offshore, the company man has an office and designated quarters. In either case, the company representative is in charge of all the operator's activities on the location. This person helps plan the strategy for drilling the well, orders the needed supplies and services, and makes on-site decisions that affect the well's progress. The company representative and the rig superintendent usually work closely together.

Area Drilling Superintendent

Large land drilling contractors, who may operate rigs all over the world and who often have several rigs working in a particular area, often employ an *area drilling superintendent*. This person's job is to manage and coordinate the activities of the many rigs the drilling company has working in a particular area or region. An area superintendent's duties include disseminating important information to each rig in the region, ensuring that all rigs are operating well and safely, and assisting each rig's superintendent when required. Area drilling superintendents frequently travel from rig to rig, so they usually have an office in a town or city in the area.

Offshore Personnel

Offshore, the sea and the remoteness of the site complicate operations. The contractor therefore requires more personnel than on land. For example, in many areas, regulations require that offshore rigs have an *offshore installation manager (OIM)*. The OIM is in charge of the entire rig and has the final say in any decision that affects the operation. In some cases, the rig superintendent is also the OIM; in other cases, the rig has an OIM as well as a rig superintendent.

Offshore contractors also hire several *roustabouts*. Roustabouts are general workers on the rig whose duties include unloading supplies from boats to the rig (fig. 48). They also keep the offshore facility in good repair. A crane operator runs the rig's cranes and supervises the roustabouts (fig. 49). Cranes transfer supplies to and from boats. Radio operators install, maintain, and repair complex radio gear that keeps the rig in constant contact with shore facilities. Medics provide first aid and are often certified emergency medical technicians (EMTs), who can stabilize injured personnel and prepare them for evacuation to shore.

On floating rigs, such as drill ships and semisubmersibles, more personnel are required because in some ways floating rigs are like ships. Not only do floating rigs drill, but also they move

Figure 48. Two rig roustabouts help move casing from a supply boat to the rig.

Figure 49. A crane operator manipulates controls from a position inside the crane cab.

on the ocean's surface just as ships do. Consequently, floaters require marine crews, individuals whose primary responsibilities have to do with the sea-going aspects of the rig.

As mentioned before, some floating offshore rigs use anchors to hold them in place on the water's surface while drilling. Other floaters employ dynamic positioning, which involves advanced computer-assisted equipment and special propellers (*thrusters*) to hold them in position on the water's surface. Such rigs require a *dynamic positioning operator*. Dynamic positioning operators maintain, repair, and monitor the equipment.

Floating rigs also require subsea equipment. Crew members place the equipment on the seafloor and operate it from the rig on the water's surface. Such equipment includes *subsea blowout preventers*. When closed, these large valves keep high-pressure fluids from escaping to the surface should the well encounter them. Accordingly, floating rigs employ *subsea equipment supervisors* (also called "subsea engineers"), whose primary job is to keep the equipment in good working order and supervise its installation on the seafloor. Often, floaters also have an assistant subsea equipment supervisor.

Figure 50. A barge engineer monitors a semisubmersible's stability from a work station on board the rig.

Also associated with floating offshore rigs are *barge engineers*, who are also called "barge masters" or "barge control operators." Semisubmersible rigs, whose pontoon-shaped hulls are submerged just below the water's surface, require barge engineers to keep the rig stable and trim while it is at work or being moved (fig. 50).

Office Personnel

Vital to any drilling project are those who work in or near company offices. Operating companies, drilling contractors, and service and supply companies hire geologists, accountants, bookkeepers, sales personnel, and trainers. They also hire personnel specialists, planners, drilling engineers, environmental specialists, warehouse personnel, and safety specialists. In addition, they employ truck drivers, storage yard personnel, lawyers, drafting technicians, and a clerical staff to back up those in the field. Without a competent office staff, no company or contractor could keep a drilling operation going. Now let's take a closer look at the stuff they're drilling for: oil and gas.

6 Oil and Gas: Characteristics and Occurrence

Oil and gas are naturally occurring *hydrocarbons*. Two elements, hydrogen and carbon, make up a hydrocarbon. Hydrogen and carbon have a strong attraction for each other. Therefore, they form many compounds. The oil industry processes and refines natural and crude hydrocarbons recovered from the earth to obtain hydrocarbon products. Products include natural gas, liquefied petroleum gas (LPG, or hydrogas), gasoline, kerosene, and diesel fuel, to name only a few.

Crude oil and natural gas occur in tiny openings in buried layers of rock. Occasionally, as at Oil Creek, nothing prevents the crude hydrocarbons from oozing to the surface in the form of a seep, or spring. More often, rock layers trap hydrocarbons thousands of feet (metres) below the surface. Operating companies and drilling contractors must therefore drill wells to bring them to the surface.

NATURAL GAS

The simplest hydrocarbon is methane (CH_4). It has one atom of carbon (C) and four atoms of hydrogen (H). Under normal pressure and normal temperature, methane is a gas. Normal pressure is the pressure the atmosphere exerts at sea level. Normal temperature ranges from about 60 to 68 degrees Fahrenheit, or 15 to 20 degrees Celsius. Methane is the main component of natural gas. Natural gas occurs in buried rock layers usually mixed with other hydrocarbon gases and liquids. Sometimes it also contains nonhydrocarbon gases and liquids such as helium, carbon dioxide, nitrogen, and water. After natural gas is produced, a gas-processing facility removes such impurities before the gas reaches consumers.

LIQUEFIED PETROLEUM GAS (LPG)

Ethane, propane, and butane often occur with natural gas. They are, however, heavier than methane so gas-processing equipment removes them from methane before the methane goes to consumers. Liquefied petroleum gas (LPG, or *hydrogas*) is mainly propane (C_3H_8) and butane (C_4H_{10}); it may also contain ethane (C_2H_6). When you compress propane and butane a little at normal temperature—that is, when you raise the pressure on them slightly above atmospheric pressure at normal temperature—they liquefy. When you release the pressure, they turn into gas. Thus, you can use LPG as a portable fuel. It travels in a pressurized container as a liquid. When you connect the container to a stove's burner, for example, LPG changes into gas when you turn on the burner and release the pressure.

CRUDE OIL

Crude oil is a hydrocarbon mixture that often occurs as a liquid, though some crude oils are very thick and dense and do not flow easily. Crude oil varies considerably in weight, viscosity, and color. It may also contain nonhydrocarbon impurities such as hydrogen sulfide. Generally, oil companies classify crude oil as light, intermediate, or heavy and, if it contains hydrogen sulfide, they call it "sour crude." Crude oil that does not contain hydrogen sulfide is classified as "sweet crude."

REFINED HYDROCARBONS

Oil refineries put crude oil through several chemical and physical processes to render it into many useful products such as gasoline, kerosene, and diesel fuel. These refined products are mixtures, or blends, of several hydrocarbons that are liquid under normal conditions. Generally, gasoline is made up of liquid hydrocarbons that are lighter in weight (less dense) than the liquid hydrocarbons that make up kerosene and diesel fuel.

OIL AND GAS RESERVOIRS

Hydrocarbons and their associated impurities occur in rock formations that are usually buried thousands of feet or metres below the surface. Scientists and engineers often call rock formations that hold hydrocarbons "reservoirs."

Oil does not flow in underground rivers or pool up in subterranean lakes, contrary to what some people think. And, as you've learned, gasoline and other refined hydrocarbons do not naturally occur in pockets under the ground, just waiting to be drilled for. Instead, crude oil and natural gas occur in buried rocks and, once produced from a well, companies have to refine the crude oil and process the natural gas into useful products. Further, not every rock can hold hydrocarbons. To serve as an oil and gas reservoir, rocks have to meet several criteria.

Characteristics of Reservoir Rocks

Nothing looks more solid than a rock. Yet, choose the right rock—say, a piece of sandstone or limestone—and look at it under a microscope. You see many tiny openings or voids. Geologists call these tiny openings "pores" (fig. 51). A rock with pores is "porous" and a porous rock has "porosity." Reservoir rocks must be porous, because hydrocarbons can occur only in pores.

A reservoir rock is also *permeable*—that is, its pores are connected (fig. 52). If hydrocarbons are in the pores of a rock, they must be able to move out of them. Unless hydrocarbons can move from pore to pore, they remain locked in place, unable to flow into a well. A suitable reservoir rock must therefore be porous, permeable, and contain enough hydrocarbons to make it economically feasible for the operating company to drill for and produce them.

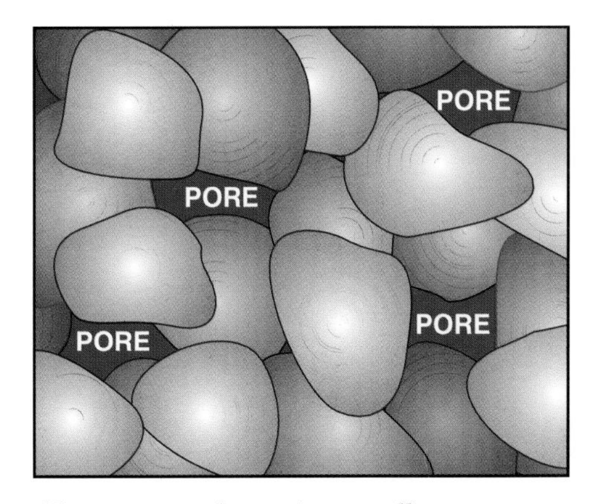

Figure 51. A pore is a small open space in a rock.

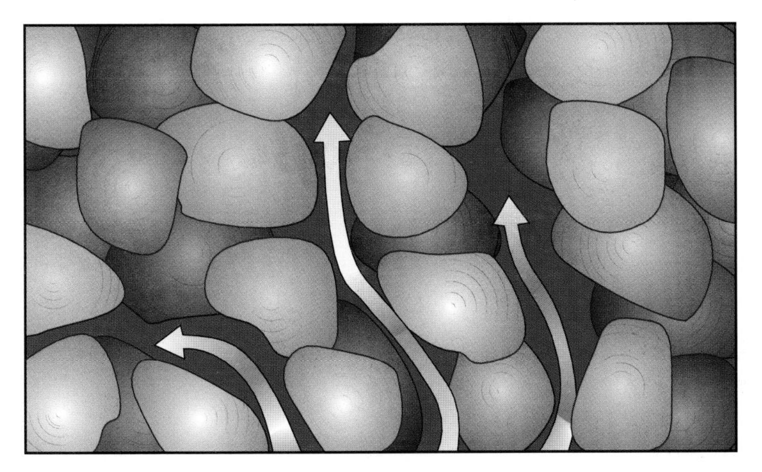

Figure 52. Connected pores give a rock permeability.

Origin and Accumulation of Oil and Gas

To understand how hydrocarbons get into buried rocks, visualize an ancient sea teeming with vast numbers of living organisms. Some are fishes and other large swimming beasts; others, however, are so small that you cannot see them without a strong magnifying glass or a microscope. Although they are small, they are very abundant. Millions and millions of them live and die daily. It is these tiny and plentiful organisms that many scientists believe gave rise to oil and gas.

When these tiny organisms died millions of years ago, their remains settled to the bottom. Though very small, as thousands of years went by, enormous quantities of this organic sediment accumulated in thick deposits on the seafloor. The organic material mixed with the mud and sand on the bottom. Ultimately, many layers of sediments built up until they became hundreds or thousands of feet (metres) thick. The tremendous weight of the overlying sediments created great pressure and heat on the deep layers. The heat and pressure changed the deep layers into rock. At the same time, heat, pressure, and other forces changed the dead organic material in the layers into hydrocarbons: crude oil and natural gas.

Meanwhile, geological action created cracks, or *faults*, in the earth's crust. Earth movement folded layers of rock upward and downward. Molten rock thrusted upward, altering the shape of the surrounding beds. Disturbances in the earth shoved great blocks of land upward, dropped them downward, and moved them sideways. Wind and water eroded formations, earthquakes buried them, and new sediments fell onto them. Land blocked a bay's access to open water, and the resulting inland sea evaporated. Great rivers carried tons of sediment; then dried up and became buried by other rocks. In short, geological forces slowly but constantly altered the very shape of the earth. These alterations in the layers of rock are important because, under the right circumstances, they can trap and store hydrocarbons.

Even while the earth changed, the weight of overlying rocks continued to push downward, forcing hydrocarbons out of their source rocks. Seeping through subsurface cracks and fissures, oozing through small connections between rock grains, the hydrocarbons moved upward. They moved until a subsurface barrier stopped them or until they reached the earth's surface, as they did at Oil Creek. Most of the hydrocarbons, however, did not reach the surface. Instead, they became trapped and stored in a layer of subsurface rock. Today, the oil industry seeks petroleum that was formed and trapped millions of years ago.

Petroleum Traps

A hydrocarbon reservoir has a distinctive shape, or configuration, that prevents the escape of hydrocarbons that migrate into it. Geologists classify reservoir shapes, or traps, into two types: *structural traps* and *stratigraphic traps*.

Structural Traps

Structural traps form because of a deformation in the rock layer that contains the hydrocarbons. Two examples of structural traps are *fault traps* and *anticlinal traps* (fig. 53).

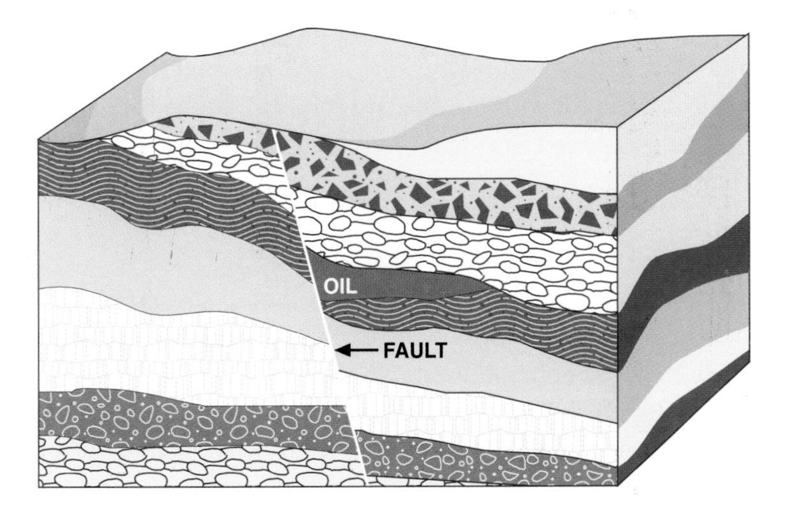

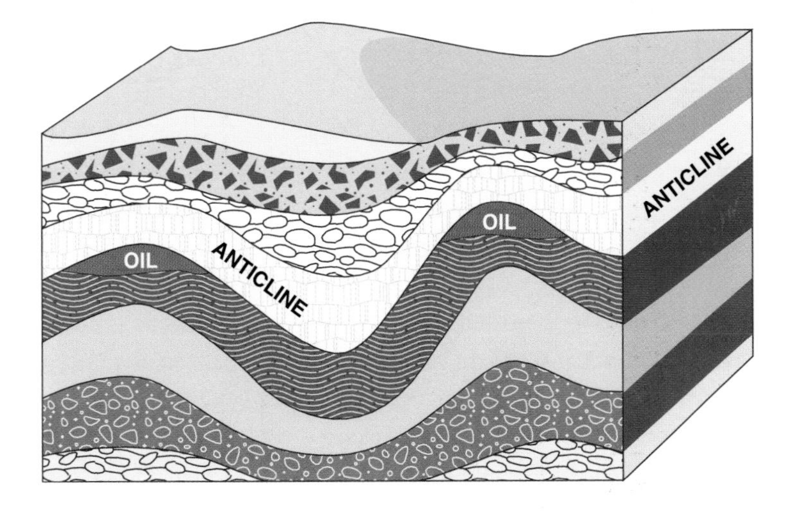

Figure 53. A fault trap and an anticlinal trap

Fault Traps

A *fault* is a break in the layers of rock. A fault trap occurs when the formations on either side of the fault move. The formations then come to rest in such a way that, when petroleum migrates into one of the formations, it becomes trapped there. Often, an impermeable formation on one side of the fault moves opposite a porous and permeable formation on the other side. The petroleum migrates into the porous and permeable formation. Once there, it cannot get out because the impervious layer at the fault line traps it.

Anticlinal Traps

An *anticline* is an upward fold in the layers of rock, much like a domed arch in a building. The oil and gas migrate into the folded porous and permeable layer and rise to the top. They cannot escape because of an overlying bed of impermeable rock.

Stratigraphic Traps

Stratigraphic traps form when other beds seal a reservoir bed or when the permeability changes within the reservoir bed itself. In one stratigraphic trap, a horizontal, impermeable rock layer cuts off, or truncates, an inclined layer of petroleum-bearing rock (fig. 54A). Sometimes a petroleum-bearing formation *pinches out*—that is, an impervious layer cuts it off (fig. 54B). Other stratigraphic traps are lens-shaped. Impervious layers surround the hydrocarbon-bearing rock (fig. 54C). Still another occurs when the porosity and permeability change within the reservoir itself. The upper reaches of the reservoir are nonporous and impermeable; the lower part is porous and permeable and contains hydrocarbons (fig. 54D).

Other Traps

Many other traps occur. In a *combination trap*, for example, more than one kind of trap forms a reservoir. A faulted anticline is an example. Several faults cut across the anticline. In some places, the faults trap oil and gas (fig. 55). Another trap is a *piercement dome*. In this case, a molten substance—salt is a common one—pierced surrounding rock beds. While molten, the moving salt deformed the horizontal beds. Later, the salt cooled and solidified and some of the deformed beds trapped oil and gas (fig. 56). Spindletop was formed by a piercement dome.

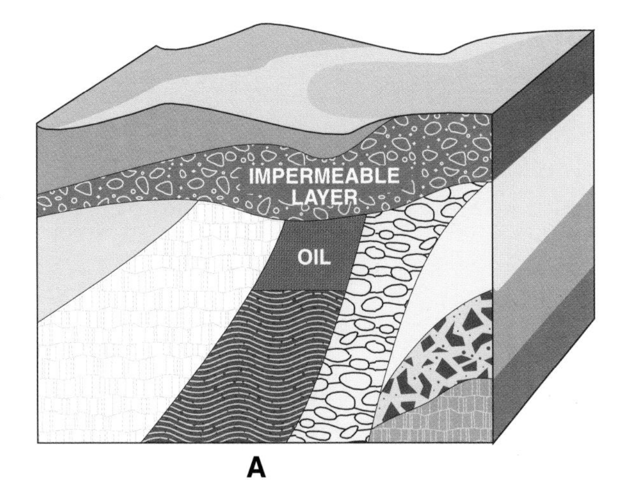

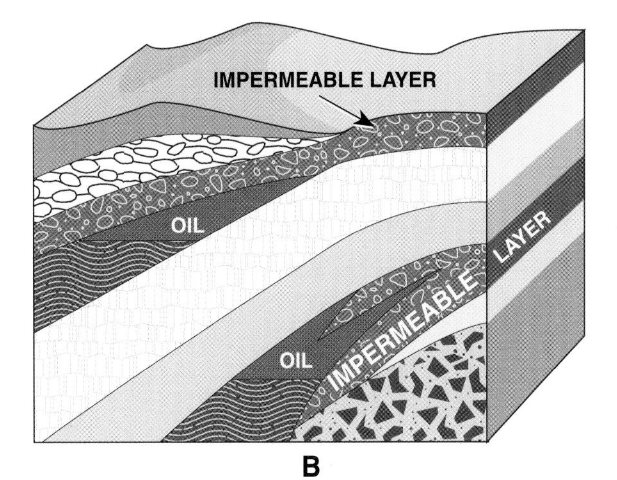

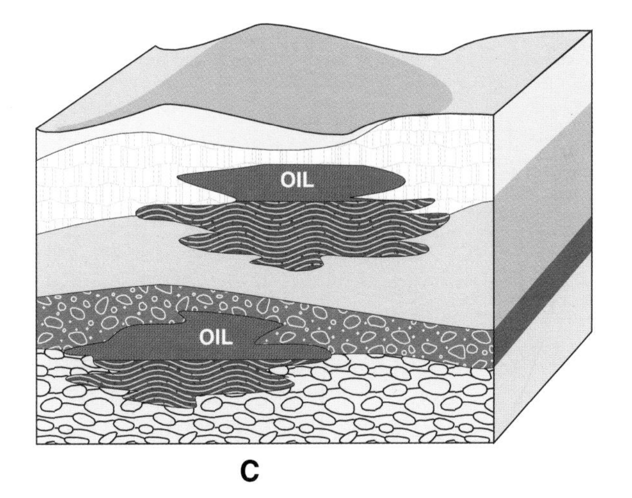

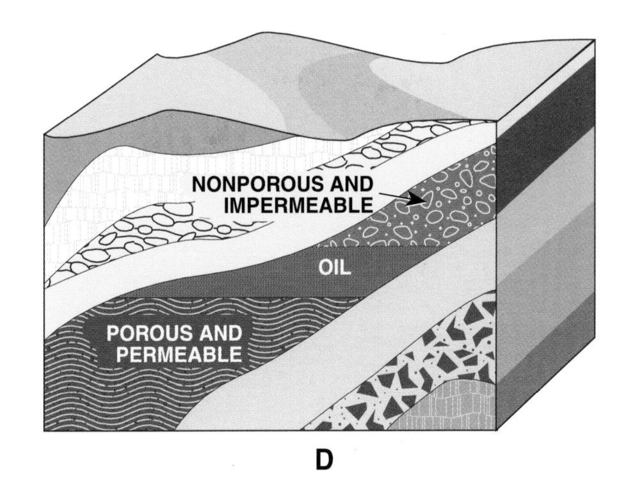

Figure 54. Stratigraphic traps

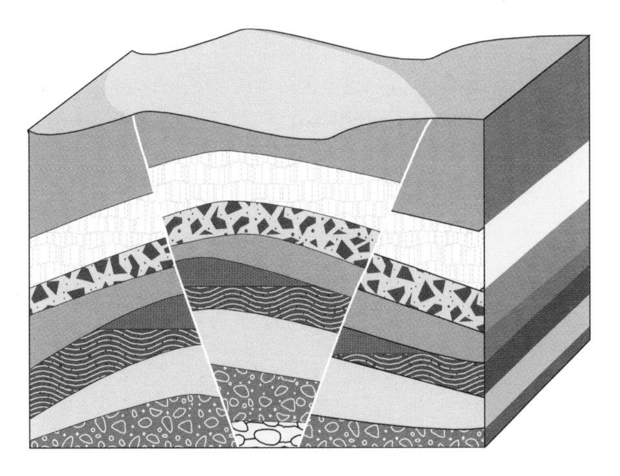

Figure 55. A faulted anticline

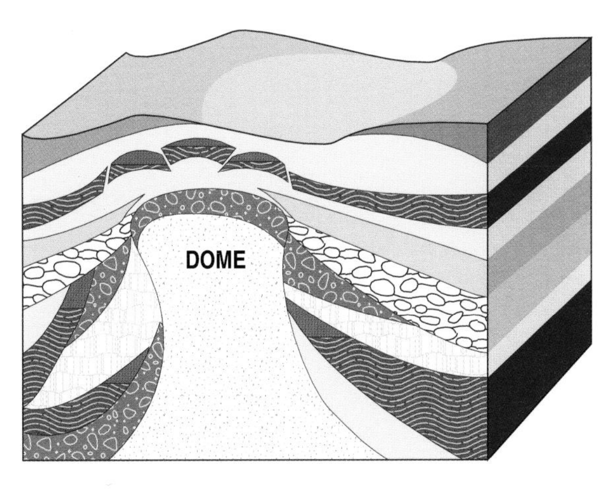

Figure 56. A piercement dome

FINDING PETROLEUM TRAPS

In the early days of oil exploration, *wildcatters* (those who drill *wildcat wells*, which are wells drilled where no oil or gas is known to exist) often drilled in an area because of a hunch. They had no idea how oil and gas occurred and probably didn't care. Anybody with enough money to back up a belief that oil lay under the ground at some location or the other drilled a well. If they were lucky, they had a strike. If not, it was on to the next hunch.

Soon, however, geologists began applying earth science to the search for oil. For example, they looked for features on the surface that indicated subsurface traps. One site was at Spindletop. An underlying salt dome created a hill, or a knoll. The knoll seemed out of place on the surrounding coastal prairie and led people like Patillo Higgins and Anthony Lucas to drill for oil.

Most petroleum deposits lie so deeply buried, however, that no surface features hint at their presence. In many places—West Texas is one example—nothing but flat, mostly featureless land stretches for many miles or kilometres. Yet, the subsurface holds large quantities of oil and gas. Consider also that much of the world's oil and gas probably lies offshore, covered by hundreds or thousands of feet or metres of water and more thousands of feet or metres of rock. Fortunately, scientists have developed effective indirect methods to view the subsurface. They use *seismology* the most.

Seismology is the study of sound waves that bounce off buried rock layers. Oil explorationists, or *geophysicists*, create a low-frequency sound on the ground or in the water. The sound can be an explosion or a vibration. If the oil hunters use explosions, the explosions create sound waves that enter the rock. If they use vibrations, a special truck forces a heavy weight against the surface and vibrates the weight (fig. 57). This vibrating weight, like an explosion, also creates sound waves that enter the layers of rock. Searchers often use several such trucks (fig. 58). Because explosions in water can kill marine life, offshore explorationists use special sound generators.

Figure 57. To the right of the tire, a large and heavy plate vibrates against the ground to create sound waves.

Figure 58. Several special trucks vibrate plates against the ground.

Figure 59. Stuck into the ground, a geophone (this one is blue) picks up reflected sound waves. Several geophones are placed in an array during a seismic survey.

Regardless of how oil seekers make the low-frequency sound, it penetrates the many layers of rock. Where one layer meets another, a boundary exists. Each boundary reflects some of the sound back to the surface. The rest continues downward. On the surface, special devices, termed "geophones," pick up the reflected sounds (fig. 59). The sounds carry information about the many layers. Cables from the geophones or hydrophones transmit the information to sophisticated recording devices in a truck or on a boat.

Explorationists take the recordings to a special laboratory where personnel use powerful computers to analyze and process the recordings. The computers display and print out the seismic signals as two- or three-dimensional views (fig. 60). Some seismic readouts show a sort of cross section of the earth. Others display a top view of buried rock layers. This type of display, in effect, removes thousands of feet of rock lying above a given layer to reveal the layer from above. Seismic displays indicate to knowledgeable personnel where oil and gas may exist. Unquestionably, seismic exploration is valuable; indeed, modern seismic technology pinpoints buried oil and gas reservoirs with great accuracy. Because of this accuracy, operating companies can be reasonably sure that when they drill a well, the reservoir it taps will produce oil or gas.

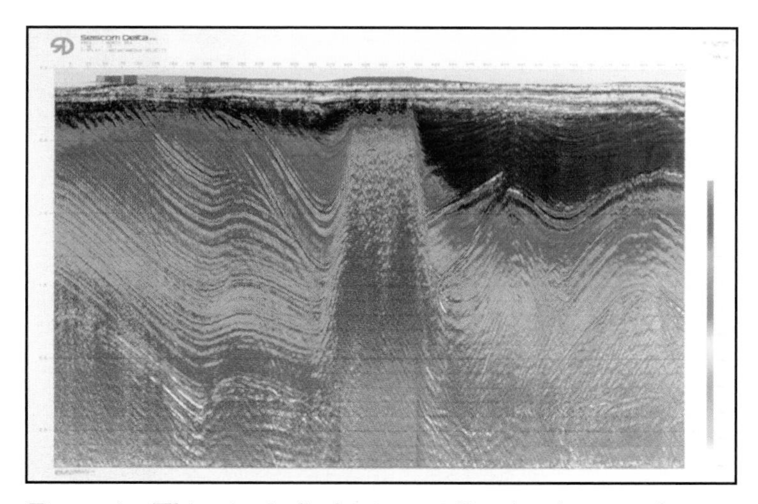

Figure 60. This seismic display is specially colored to reveal details of earth formations.

TYPES OF WELLS

The industry generally classifies wells as "exploration wells," "confirmation wells," and "development wells." They also speak of drilling "infilling" wells and "step-out" wells. An exploration, or wildcat, well is one a company drills to determine whether oil or gas exists in a subsurface rock formation. A wildcat well thus probes the earth where no known hydrocarbons exist.

If a wildcat well discovers oil or gas, the company may drill several confirmation wells to verify whether the wildcat well tapped a rock layer with enough hydrocarbons for the company to develop it. One well that finds oil and gas may not produce enough to justify the company's preparing it for production. Usually, several wells must produce for the company to get its money back and to make a profit.

A development well is drilled in an existing oilfield. A company drills this type of well so it can remove more hydrocarbons from the field. Engineers carefully study a field's producing characteristics. They then determine the number of wells required to produce it efficiently. They sometimes call development wells "infilling wells" if a contractor drills them between existing wells. If the company drills wells on the edge of an existing field, perhaps to determine the field's boundaries, they may call them "step-out wells," or "outpost wells."

The number of development wells drilled into a particular reservoir depends mainly on its size and characteristics. A reservoir can cover several acres (hectares) and may be only a few feet (metres) thick or hundreds of feet (metres) thick. In general, the larger the reservoir, the more wells it takes to produce it. Reservoir characteristics, such as porosity and permeability, also play a role. For example, a reservoir with high porosity and permeability, which allow the hydrocarbons to flow easily, may not require as many wells to produce as a reservoir with low porosity and permeability.

Regardless of the type of well, before the drilling contractor can drill it, the operator must make the drill site ready. So, let's next examine drill site preparation.

The Drill Site

7

The *drill site*—the location of the well—varies as the surface geography of the earth varies. In the early days of the industry, geologists and wildcatters were able to find oil and gas in places that were generally accessible. As people began to use more hydrocarbons, however, the oil industry extended its search for oil and gas to all corners of the globe. Today, companies drill wells in frozen wildernesses, remote deserts, mosquito-ridden marshes, hot and humid jungles, high and rugged mountains, and deep offshore waters. In short, a drill site is anywhere oil and gas exist or may exist.

CHOOSING THE SITE

The operating company decides where to drill by considering several factors. The most important is that the company knows or believes that hydrocarbons exist in the rocks beneath the site. In some cases, the operator drills a well in an existing field to increase production from it. In other cases, the operator drills a well on a site where no one has found oil or gas before. The company often hires geologists to find promising sites where no production exists. Geologists explore areas to try to determine where hydrocarbons may exist. Major companies sometimes have a staff of geologists; independents often hire consulting geologists or buy information from a company that specializes in geological data.

Legal and economic factors are also important in the selection of a drilling site. For example, the company must obtain the legal right to drill for and produce oil and gas on a particular piece of land. Further, the company must have money to purchase or lease the right to drill and produce. What's more, it must have money to pay for the costs of drilling. The costs of obtaining a lease and drilling for oil or gas on the lease vary considerably. Costs depend on such factors as the size of the reservoir, its depth, and its location (offshore and remote sites cost more to drill and produce than readily accessible land sites). A company can easily commit several million dollars to find, drill for, and produce oil and gas. The rewards, of course, can be great, but so can the expenses.

The operating company takes several steps before telling the drilling contractor exactly where to place the rig and start, or *spud*, the hole. The company carefully reviews and analyzes seismic records. Legal experts thoroughly examine lease terms and agreements. They ensure that the operating company has clear title and right-of-way to the site. Surveyors establish and verify exact boundaries and locations. The company also confirms that it has budgeted the necessary drilling funds and that the funds are available.

On land, operating personnel usually try to choose a spot directly over the reservoir. With luck, the surface will be accessible and reasonably level. They also try to pick a location that will not suffer too much damage when the contractor moves in the rig. In an area that is especially sensitive, the operator and contractor take extra steps to ensure that as little harm as possible occurs. Offshore, the operator hopes that the weather is reasonably good, and, if using a bottom-supported rig, picks a spot where the ocean bottom (the *mud line*) can adequately hold any rig supports in contact with it.

PREPARING THE SITE

On land sites, the operator hires a site-preparation contractor to prepare the location to accommodate the rig. If required, bulldozers clear and level the area. This contractor also builds an access road and, if necessary, a turnaround. Offshore, the operator simply marks the spot with a buoy. On all jobs, contractors and operators make every effort to keep damage to a minimum because no one wishes to harm the environment. Further, if harm does occur, the contractor and operator have to pay to correct or mitigate the damage, which can be expensive.

Surface Preparation

The contractor uses various materials to prepare the surface and roads around a land location. Near the coast, oyster shells are popular. In other locations, gravel may be the choice. A contractor may lay boards to allow access in rainy weather. In the far north, permafrost presents a special problem because the heat generated under and near the rig may melt the permafrost. Thus, the rig may settle into the thawed soil. In permafrost, therefore, the contractor spreads a thick layer of gravel to insulate the area. If gravel is scarce, polyurethane foam may be used.

Reserve Pits

At a land site, the site-preparation contractor may dig a *reserve pit*. A reserve pit is an open pit that is bulldozed from the land next to the rig. Reserve pits vary in size, depending on how much room is available at the site. Usually, reserve pits are relatively shallow, maybe no more than 10 feet (3 metres) deep and are open on top. In the early days of drilling, the reserve pit was mainly a place to store a reserve supply of drilling mud. Today, however, drilling mud used in actively drilling the hole is seldom stored in the reserve pit, although, in an emergency, it can be.

Modern reserve pits mainly hold rig wastes temporarily. For example, cuttings carried up the hole by the drilling mud fall into the reserve pit. After finishing the well, the drilling contractor or operator removes any harmful material that may be in the pit and properly disposes of it. A bulldozer then covers it with dirt and levels it. If necessary, the contractor lines a reserve pit with plastic to prevent soil and groundwater pollution (fig. 61). In especially sensitive areas, such as in a migratory bird flyway or in a wildlife refuge, contractors cover the pit with netting to prevent birds from landing in it. In addition, they may put up a fence to keep cattle or wildlife out.

Figure 61. A reserve pit

In some areas, reserve pits are rare. Offshore, and on sensitive land locations, the contractor places cuttings in portable receptacles and disposes of them at an approved site. Most operators and contractors recycle as many drilling mud components and other materials as possible. What they cannot recycle, they discard at approved sites. In some areas, regulatory agencies enforce a zero-discharge policy. This policy prevents anyone from emitting anything onto the ground, into a waterway or estuary, or into the sea.

Cellars

The operator may make additional preparations before moving in the rig. The terrain, the well's depth, the underground pressures expected, and the operator's and contractor's preferences determine how they start the well. At land sites where the operator has ordered a deep, high-pressure well, for example, a work crew, using dirt moving equipment, may dig a rectangular pit, or *cellar*. Sizes vary, but a typical cellar is about 10 feet (3 metres) on a side and perhaps 10 feet (3 metres) deep. The exact size and depth depend on the characteristics of the well and the rig's configuration.

Sometimes, the workers line the cellar with boards or pour concrete walls to keep it from caving in (fig. 62). The cellar accommodates a tall stack of high-pressure control valves under the rig. The bottom of the stack will sit in the cellar, below ground level. Since the crew installs the stack below ground level, the rig's *substructure*—the base of the rig—does not have to be as tall to allow the rig floor to clear the stack. In short, a cellar provides more working room under the rig.

Figure 62. A concrete pad surrounds the cellar at this drilling site; note the large pipe (with a temporary cover) protruding from the middle of the cellar. The rig will drill into this pipe to start the well. The device to the right of the cellar contains valves that crew members will install on top of the well to control pressure.

Rathole

Some rigs use a special pipe called the "kelly," which is part of the drill string. The kelly is part of the system that rotates the bit. Rigs with kellys require a *rathole*—a shallow hole drilled off to the side of the main borehole. On land, the operator sometimes hires a special truck-mounted, light-duty unit called a "rathole rig" to drill the rathole. Or, after the rig is set up (*rigged up*), the drilling crew may drill the rathole with special equipment. Offshore, if the rig needs a rathole, it is a large-diameter length of pipe that extends below the rig floor. In the case of drilled ratholes, the crew extends pipe from the drilled part of the rathole up to the rig floor. The rathole goes through the rig floor and protrudes a few feet, or a half metre or so, above it (fig. 63).

Figure 63. Part of the rig's drilling assembly is temporarily placed in the rathole.

During drilling, the crew uses the rathole to store the kelly temporarily. A kelly can be up to 54 feet, or 17 metres, long. Even the tallest land rig substructures are only about 40 feet (12 metres) high and most are even shorter. The contractor therefore has to drill part of the rathole; otherwise, the rathole would extend too high above the rig floor to be accessible.

Mousehole

The rathole rig or the main rig itself may also drill a *mousehole* on land sites. A mousehole, like a rathole, is also a shallow hole lined with pipe that extends to the rig floor. The mousehole is a lined hole into which the crew puts a length, or *joint*, of drill pipe during drilling operations (fig. 64). When crew members are ready to add the joint to the drill string as the hole deepens, they add it from the mousehole. A joint of drill pipe is around 30 feet (9 metres) long. If the regular rig's substructure is appreciably shorter than this height, then the rathole crew also drills a mousehole.

Figure 64. A joint of drill pipe rests in this rig's mousehole.

Figure 65. A rathole rig drills the first part of the hole.

Conductor Hole

The rathole crew may also drill the first, or top, part of the main borehole. The operator can, in some cases, save time and money by having the rathole rig actually start, or spud, the main hole before moving in the regular rig. The rathole crew backs the rathole rig to the cellar (fig. 65). A special bit starts the main hole in the middle of the cellar. This hole is shallow in depth but large in diameter. Termed conductor hole, it may be 36 inches (91 centimetres) or more in diameter (fig. 66). It may be only tens of feet (or metres) deep or it may be hundreds of feet (or metres) deep, depending on the surface conditions.

Figure 66. The conductor hole

Figure 67. The large diameter pipe to the right is the top of the conductor pipe. The small diameter pipe to the left lines the rathole. Later, the rig crew will install a tube to extend the rathole up to the rig floor.

The rathole crew lines the conductor hole in the cellar with conductor pipe. Conductor pipe, or casing, keeps the hole from caving in. It also conducts drilling mud back to the surface when regular drilling begins (fig. 67). The crew often secures the conductor pipe in the hole with cement or concrete. With the conductor pipe, rathole, and mousehole prepared, the drilling contractor can move in the rotary rig to drill the rest of the hole.

Other Considerations

On drilling locations where the ground is soft, a rathole rig and crew may not be needed. Instead, the contractor can usually move in the regular rig and its crews right away. Once the drilling crew members get the regular rig ready, they rig up a pile driver and drive the conductor casing into the ground, just as Colonel Drake did at Oil Creek. Thus, people in the oil patch sometimes call conductor casing "drive pipe." After driving the casing, the rig crew begins drilling inside it.

If the ground is too hard for the conductor pipe to be driven, crew members can use the regular rig to drill the conductor hole. What's more, they may also drill the rathole and mousehole, using special equipment on the regular rig.

MOVING EQUIPMENT TO THE SITE

After the operator selects and prepares the drill site, the contractor moves the rig to the site. Crew members move most land rigs by loading the rig components onto trucks. The trucks then carry the components to the site where crew members put the components back together and begin drilling. In remote areas, such as in jungles and arctic regions, crew members may load rig components onto cargo airplanes or helicopters. Boats often tow offshore rigs from one site to another. On the other hand, some offshore rigs are self-propelled—that is, built-in units on the rig provide the means to move it. Sometimes, especially where a rig has to be transported a long distance, a special ship carries the rig.

Moving Land Rigs

Virtually all land drilling rigs are portable. If the rig is small enough to be built on a truck, a person simply drives it from one place to another. Once at the site, the rig stays on the truck and drilling commences. Rigs too big to fit onto one truck are designed differently. Fabricators design medium and large rigs so that a contractor's crew can take it apart, load its components onto several trucks, helicopters, or cargo planes, and move it to the drilling site. At the site, crew members put the rig together, or *rig up*. After they drill the well, they dismantle the rig, or *rig down*.

As mentioned earlier, in deserts and other flat places, the contractor may skid the rig. A rig suitable for skidding has enormous wheels attached to the substructure, which, when engaged, allow the rig to be towed short distances without a crew's having to dismantle it.

Moving and Setting Up Offshore Rigs

Some offshore rigs are self-propelled. Built-in engines and screws (propellers) move the rig through the water. Rudders like those on a ship allow marine personnel to steer the rig when it is underway. While a self-propelled rig's speed is slow—perhaps 3 or 4 knots per hour at the fastest—generally, the distances traveled are relatively short, so speed is not a factor. For rigs that are not self-propelled, the contractor can hire boats to tow them.

For long moves, say from one ocean to another, the contractor may use a special ship to carry the rig, whether it is self-propelled or not (fig. 68). To load the rig onto the ship crew members moor the ship next to the rig, usually in the shallow waters of a port. At first, both the boat and the rig float. They then flood compartments in the ship to submerge its deck below the waterline. With the deck below the water's surface, large cranes pull the rig over to the ship's deck. Pumps remove the water from the compartments and the ship floats back to the water's surface with the rig in place on the deck.

Whether on land or offshore, once the site is prepared for the rig, the next step is for the drilling crew to *rig up*—that is, to put the rig components together and prepare the rig for drilling. So, let's look next at rigging up.

Figure 68. A special ship carries a semisubmersible to a new drilling location.

Rigging Up

8

Rigging up an offshore drilling rig is usually not as complicated as rigging up a land rig. Most offshore rigs can be moved across the water's surface with virtually no disassembly of major parts. To move most land rigs, on the other hand, crew members must disassemble many of its components. Disassembly is required so they can load the parts on trucks, planes, or helicopters for transportation to the next location. Once the contractor gets the land rig to the site, the next step is for the drilling crew to put the rig together, or to *rig up*. For safety's sake, rig up usually occurs only during daylight hours. A rig usually has too much heavy equipment moving around during rig up for it to be safe in the dark, even with lights.

For most land rigs, rigging up means to put the rig parts back together so that the rig can drill a hole. It involves unloading and hooking up the rig engines, the mud tanks and pumps, and other equipment on the site. One of the last steps, and one of the most dramatic, is raising the mast from horizontal—the position it was transported in—to vertical to ready it for drilling. After unloading and hooking up the engines to get power, crew members position the rig's *substructure*, which is its base, or foundation.

SUBSTRUCTURES

A substructure is a framework that rests directly over the hole and is the foundation of the rig. The bottom of the substructure rests on level ground. Crew members place a work platform on top of the substructure that they term the "rig floor." The substructure raises the rig floor from about 10 to 40 feet (3 to 12 metres) above the ground. Elevating the rig floor makes room under the rig for the special high-pressure valves and other equipment that the crew connects to the top of the well's casing. The exact height of a substructure depends on the space needed to clear this equipment. Remember, too, that, in some cases, a cellar provides more space for the equipment.

Figure 69. A box-on-box substructure

One type of substructure is the box-on-box (fig. 69). Using trucks or portable cranes, the crew stacks one steel-frame box on top of another to achieve the desired height. Another, more modern substructure is the *self-elevating*, or *slingshot*, type. Crew members place it on the site in a folded position (fig. 70). They then activate winches to unfold and raise the substructure to full height (fig. 71). Slingshot substructures go up much faster than the box-on-box type. Whether box on box or slingshot, the substructure is rugged because it not only supports the weight of the drilling equipment that the crew assembles on top of it, but also the weight of the entire drill string.

Figure 70. A slingshot substructure in folded position prior to being raised; the rectangular structure resting on the folded substructure is the rig's doghouse, which is a sort of driller's office on the rig floor.

Figure 71. The slingshot substructure near its full height

Figure 72. This drawworks will be installed on the rig floor. Some are installed below the rig floor.

Figure 73. Drilling line is spooled onto the drawworks drum.

RIG FLOOR

Crew members set many pieces of equipment on the substructure, including a steel-and-wood rig floor on which to work. An important piece of equipment that the rig floor may support is the *drawworks* (fig. 72). (Some rig designs call for the drawworks to be installed below the rig floor.) Whether at floor level or below, the drawworks is a large hoist. It houses a *spool*, or a *drum*, on which the crew wraps braided steel cable called "drilling line" (fig. 73). The drilling line is large-diameter wire rope, which ranges in size from ⅞ to 1½ inches (22.23 to 38.1 millimetres). It has to be big because it carries and moves an incredible amount of weight as the well is drilled. The drawworks, the drilling line, and the mast or derrick support practically everything that goes in or comes out of the hole.

Figure 74. A mast being raised to vertical position

RAISING THE MAST OR DERRICK

Crew members next raise the mast or derrick. If the rig has a mast, they raise it from horizontal to vertical with the drawworks (fig. 74). If the rig has a standard derrick, crew members bolt it together, one piece at a time, on the substructure. After finishing the well, they have to disassemble the derrick and rebuild it at a new site. When assembled, a standard derrick looks very much like a mast. Unlike a mast, however, which the crew raises or lowers as a complete unit from a pivot point in the substructure, a derrick has four legs that extend from each corner of the substructure (fig. 75). Virtually every land rig today uses a mast. Occasionally, you may find a standard derrick on an offshore platform.

Figure 75. This rig with a standard derrick was photographed in the 1970s at work in West Texas. The rig crew erected it piece by piece. After completing the well, they took it down piece by piece and moved it to another location.

Derricks get their name from a hangman who plied his trade in Tyburn, England (today, a suburb of London), in the 1600s. Even though Derick's first name is unknown, his gallows were so distinctive that everyone began calling any towerlike structure with cross braces and girders a "derick." By the nineteenth century, scholars had added the extra "R."

Crew members can raise or lower a mast without completely assembling and disassembling it each time the rig moves. Not having to build and take apart a derrick is a timesaving advantage. Once the manufacturer constructs all the braces, girders, and cross members of a mast, no one totally disassembles it again until the contractor scraps it. Some masts fold, telescope, or even come apart into sections to make them shorter and easier to move. Nevertheless, they retain their integrity as a unitized component.

Derrick and Mast Heights

A derrick or mast is a tall and strong tower that supports the entire weight of the drill string and other tools that the crew runs in and out of the hole. The drill string may be thousands of feet long. To pull the string, the crew has to unscrew it, or *break it out*, into smaller lengths, or *stands*. Usually, crew members set each stand back in the derrick or mast in the vertical position after they pull it. The tallest derricks and masts are about 200 feet, or 60 metres, high. The shortest are about 65 feet, or 20 metres, high.

As mentioned before, a derrick (usually called a "standard derrick") is a structural tower that a crew assembles piece by piece. Many people in the drilling business, however, call a mast a derrick regardless of what it really is. The proper term, however, is "mast," so this book uses mast when referring to a mast and derrick when referring to a standard derrick. It uses mast or derrick when it could be either.

Mast or derrick height governs whether the crew pulls pipe from the hole in *singles*, *doubles*, *triples*, or *quadruples*. (Old drilling hands called triples, "thribbles" and quadruples, "fourbles." You may still hear these terms applied to pulling pipe in three- or four-joint stands.) If crew members "pull singles," they break out the pipe a single joint at a time. A single length, or joint, of drill pipe is about 30 feet, or 9 metres, long. If they pull doubles, they break out the pipe in two-joint stands of about 60 feet, or 18 metres. If they pull triples (thribbles), they break out the pipe in three-joint stands of about 90 feet, or 27 metres. If they pull quadruples (fourbles), they break out the pipe in four-joint stands of about 120 feet, or 36 metres. In any case, the longer the stand, the faster the crew can pull the pipe and return it to the hole.

Mast Load Ratings

Masts have to be strong; at the same time, they have to be portable. Manufacturers rate masts by the vertical load they can carry and by the amount of wind they can withstand. Mast capacities for vertical loads run from 0.25 million up to 1.5 million pounds (over 0.5 million to about 3 million kilograms). In some cases, the drill string alone may weigh as much as a half million pounds, or over one million kilograms. Most masts can withstand winds of 100 to 130 miles (160 to 210 kilometres) per hour.

RIGGING UP ADDITIONAL EQUIPMENT

Rig up involves the assembly of lots of equipment. Crew members install large engines to power the rig. They set up steel tanks in which they put drilling mud. They also connect the pumps that will move the mud down the hole. They erect safe stairways and walkways to allow access to the many components. They position auxiliary equipment for generating electricity, compressing air, pumping hydraulic fluid, and pumping water.

Truck drivers and *swampers*, their helpers, bring in storage racks, bins, and living quarters for the company representative and the rig superintendent. They also deliver drill pipe, pipe racks, fuel tanks, wire rope, and other items to the location. Crew members place a small metal building called a doghouse adjacent to the rig floor (fig. 76). The doghouse is a kind of office for the driller and the crew. They keep small tools and current drilling records in the doghouse, and each crew member may have a locker for clothes and food.

Figure 76. The doghouse is usually located at rig-floor level.

Within a few hours or days, depending on the size and complexity of the rig, the crew has the rig assembled and is ready to *break tour*—that is, to begin operating 24 hours a day. (On a drilling rig, whenever crew members begin an operation, they often say they "break it." For example, a driller *breaks circulation* by turning on a mud pump to begin circulating drilling mud. To distinguish between something being started and something being broken in the usual sense of the word, you have to know the context.) As mentioned earlier, rig up usually occurs only during daylight. When the crews have the rig ready for operation, however, they begin working three 8-hour tours or two 12-hour tours.

OFFSHORE RIG-UP

Rigging-up operations offshore vary with the type of rig. To rig up a platform, for example, a construction crew builds it. After crew members pin the jacket to the seafloor, they set up the living quarters and drilling equipment on top of the jacket. They then add such items as mud tanks, a helicopter deck, and cranes. Mobile offshore rigs require less rig-up time than platforms because most of the equipment is already in place and assembled. Once the crew gets a floating rig anchored or dynamically positioned on the site, drilling operations can begin. Jackups are ready to operate after the crew jacks the legs into contact with the seafloor and raises the drilling deck above the waterline.

But, whether the rig works offshore or on land, once crew members finish rigging up, they break tour and begin drilling. To successfully drill, a rig requires many pieces of equipment. So, let's next examine the major components of a rotary drilling rig.

9 Rig Components

The main function of a rotary rig is to drill a hole, or as they say in the oil patch, to "make hole." Making hole with a rotary rig requires qualified personnel and a lot of equipment. To learn about the components it takes to make hole, let's divide them into four main systems: power, hoisting, rotating, and circulating.

POWER SYSTEM

Every rig needs a source of power to run the hoisting, circulating, and, in many cases, the rotating equipment required to make hole. In the early days of drilling, steam engines powered most rigs. For example, Colonel Drake powered his rig with a wood-fired steamboat engine (see fig. 12). Until the 1940s and 50s, steam engines drove almost every rig (fig. 77). But, as powerful and portable diesel and gas engines became available, mechanical rigs began to supplant steam rigs (fig. 78). Oil people called them "mechanical rigs" or "power rigs," because the engines drove special machinery, which, in turn, provided power to the components. Then, in the 1970s and 80s, electric generators, driven by diesel engines, began to replace the mechanical equipment used to drive rig components. Today, these "electric rigs" or "diesel-electric rigs" dominate the drilling scene.

Figure 77. In the foreground is a coal-fired boiler that made steam to power the cable-tool rig in the background.

Figure 78. A mechanical rig drilling in West Texas in the 1960s; rigs like this still drill in many parts of the world.

Whether mechanical or electric, virtually every modern drilling rig uses internal-combustion engines as a prime power source, or *prime mover*. A rig's engines are similar to the one in a car except that rig engines are bigger, more powerful, and do not use gasoline as a fuel. What's more, most rigs require more than one engine to furnish the needed power (fig. 79). Most rig engines today are diesels, because diesel fuel is safer to transport and store than other fuels such as natural gas, LPG, or gasoline.

Figure 79. Three diesel engines power this rig.

Diesel engines do not have spark plugs as do gasoline engines. Instead, heat generated by compression ignites the gaseous fuel-air mixture inside the engine. Anytime a gas is compressed, its temperature rises. Compress it enough (as in a diesel engine), and, if the gas is flammable, it gets hot enough to ignite. Thus, diesel engines are sometimes called "compression-ignition engines." Gasoline engines are often called "spark-ignition engines."

A rig, depending on its size and how deep a hole it must drill, may have from one to four engines. Naturally, the bigger the rig, the deeper it can drill and the more power it needs. Thus, big rigs have three or four engines, all of them together developing up to 3,000 or more horsepower (2,100 kilowatts). Of course, once the engines develop all this power, it must be sent or transmitted to other rig components to make them work. Electric generators transmit power on most rigs. However, a few older rigs use machinery to transmit the power.

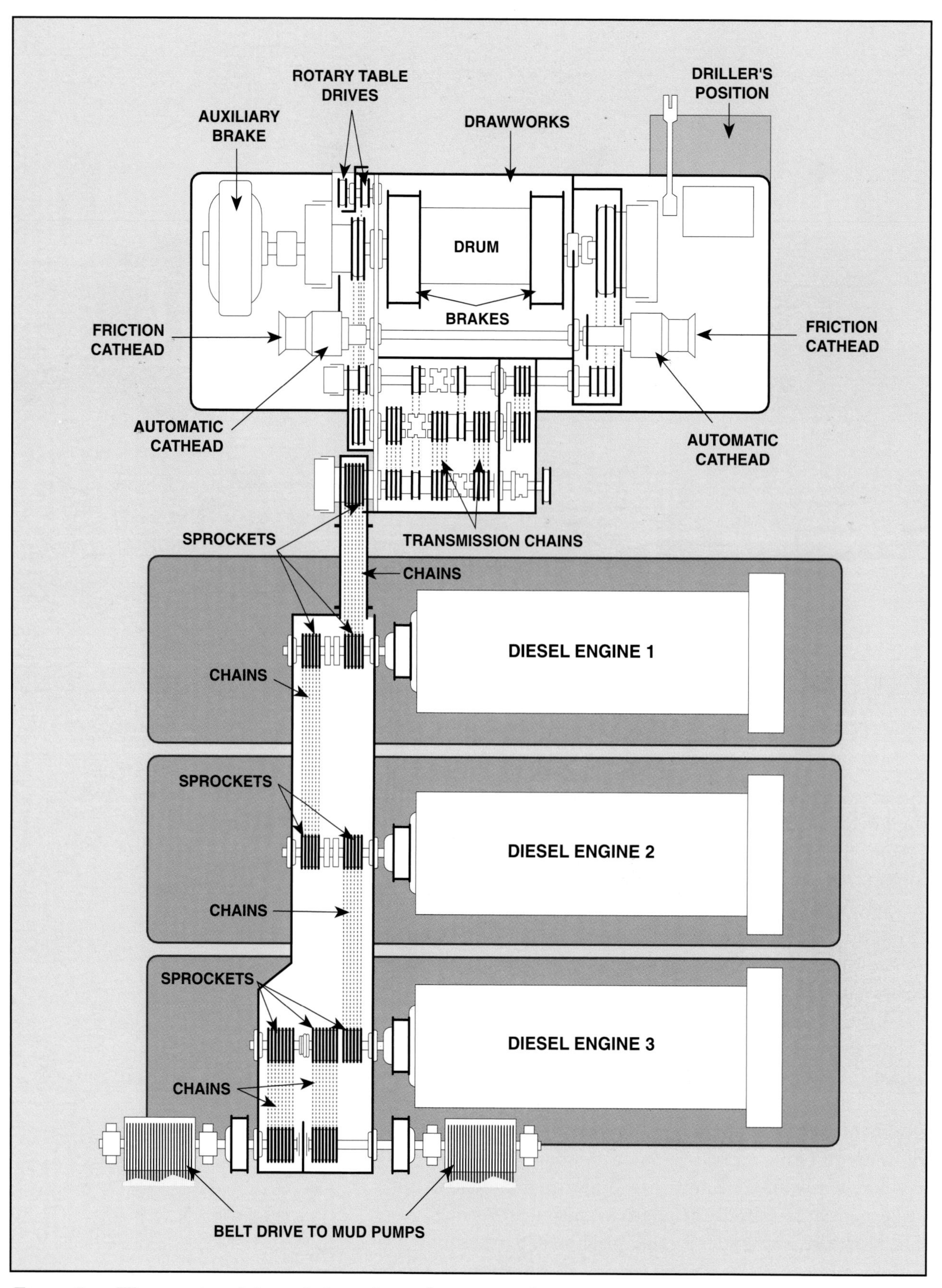

Figure 80. Three engines drive a chain-and-sprocket compound to power equipment.

Mechanical Power Transmission

As mentioned before, during the 1950s and 60s, most rigs were mechanical—that is, the engines drove big chains and sprockets, which, in turn, powered various parts of the rig. Later, diesel-electric rigs began to dominate the scene because electrical power transmission had so many advantages over mechanical transmission. Thus, today, the majority of rigs are diesel-electric. But, mechanical rigs are still around, so it's worth a brief look at mechanical power transmission.

Figure 80 is a schematic of a mechanical rig. It shows three 700-horsepower (490-kilowatt) engines hooked up to a *compound*. The compound consists of several heavy-duty sprockets and chains. The engines drive the sprockets around which the chains are wrapped. The chains drive the various rig components. This chain-and-sprocket arrangement is known as the compound because it compounds or connects the power of several engines. With compounded engines, the driller can use one, two, or all of them at once if required.

Electrical Power Transmission

On diesel-electric rigs, powerful diesel engines drive large electric generators (fig. 81). The generators, in turn, produce electricity that flows through cables to electric switch and control equipment enclosed in a control cabinet (fig. 82).

Figure 81. The diesel engine at right drives the electric generator attached directly to the engine. This engine-generator set is one of three on this rig.

Figure 82. The central control cabinet of a diesel-electric rig

From the control gear, electricity goes through more cables to electric motors. The manufacturer attaches the motors directly to the equipment to be driven—for example, the drawworks or mud pumps (fig. 83).

The diesel-electric system has a number of advantages over the mechanical system. The diesel-electric system eliminates all the heavy and fairly complicated machinery making up the compound. Because an electric rig does not require a compound, crew members do not have to spend time lining up and connecting the compound with the engines and drawworks. Also, on an electric land rig, the rig designer can position the engines well away from the rig floor so that crew members enjoy less engine noise. On mechanical rigs, the compound-to-engine set up requires that the engines be placed close to the rig floor.

Figure 83. Two powerful electric motors drive the drawworks on this rig.

HOISTING SYSTEM

Whether mechanical or diesel-electric, a rig's job is to drill a hole; to do this job it must have a hoisting system (fig. 84). A typical hoisting system is made up of the drawworks (or hoist), a mast or derrick, the crown block, the traveling block, and the wire rope drilling line.

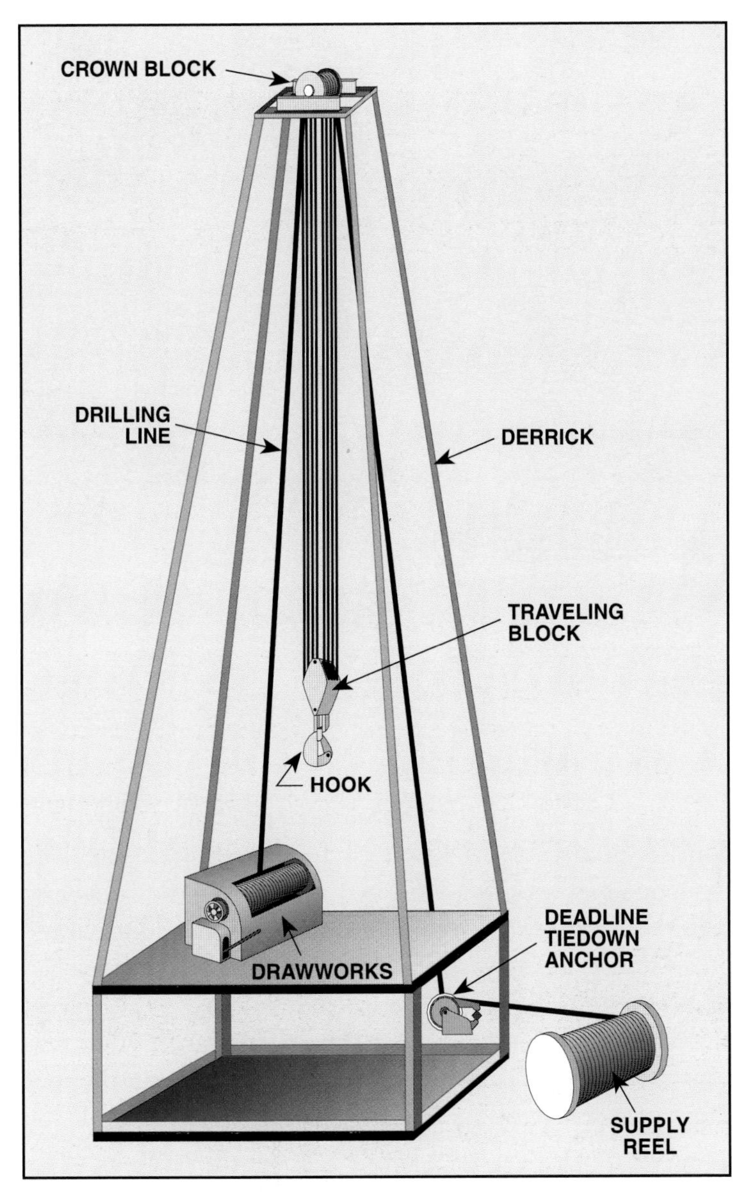

Figure 84. The hoisting system

Figure 85. The drawworks

The Drawworks

The drawworks is a big, heavy piece of machinery (fig. 85). It consists of a revolving drum around which crew members wrap or spool wire rope drilling line. It also has a *catshaft* on which the *catheads* are mounted. Further, it has clutches and chain-and-gear drives so that the driller can change its speed and direction. A main brake provides the driller a way of slowing and stopping the drum (fig. 86). An auxiliary electric brake (fig. 87) assists the main brake by absorbing the momentum created by the load being raised or lowered. Big electromagnets inside the auxiliary brake oppose the turning forces on the drum and help the main brake stop the load.

Figure 86. Removing the drawworks housing reveals the main brake bands to the left and right on hubs of the drawworks drum.

Figure 87. The electromagnetic brake is mounted on the end of the drawworks. It helps the main brake slow and stop the drawworks drum.

Figure 88. In this photo taken in the 1960s, a floorhand has fiber rope (a catline) wrapped around a friction cathead to lift an object on the rig floor.

The Catheads

A *cathead* is a winch, or windlass, on which a line, such as rope, cable, or chain is coiled. When activated, a cathead reels in the line with great force. Pulling on a line is vital to screwing and unscrewing (*making up* and *breaking out*) drill pipe. Typically, four catheads are mounted on the *catshaft* of the drawworks, two on each end. On the very ends of the catshaft are friction catheads. Right next to them are the automatic or mechanical catheads (see fig. 80).

A friction cathead is a steel spool a foot (30 centimetres) or so in diameter. It revolves as the catshaft revolves. In the old days, crew members employed friction catheads to move heavy equipment around the rig floor. One floorhand rigged up one end of the catline to the object they wished to move. Another wrapped the other end of the catline around the cathead and used it to lift the object (fig. 88). Today, rig personnel use a small, air-powered hoist to move equipment on and around the rig floor (fig. 89). They are separate from the drawworks and are called *air hoists* or *air tuggers*. Air hoists are so much easier to use than a friction cathead that, today, about the only time you might see a friction cathead being used is in an emergency.

While crew members seldom use friction catheads, they use mechanical or automatic catheads a great deal. They employ an automatic cathead to make up or break out the drill string when running it into or pulling it from the hole. It is called an automatic cathead because the driller simply moves a control to engage or disengage it. When engaged, an automatic cathead pulls on a wire rope or, in some cases, a chain, to make up or break out the string.

The automatic cathead on the driller's side of the drawworks is the *makeup cathead* because it plays a part when the crew makes up drill pipe (fig. 90). The automatic cathead on the other side of the drawworks is the *breakout cathead* because the driller engages it to break out drill pipe.

Figure 89. This floorhand is using an air hoist, or tugger, to lift an object.

Figure 90. This makeup cathead has a chain coming out of it that is connected to the tongs. When actuated, the cathead pulls on the chain to apply tightening force to the tongs. Note the unused friction cathead; its paint shows no signs of wear.

Figure 91. Wire rope drilling line coming off the drawworks drum

The Blocks and Drilling Line

Manufacturers make drilling line from very strong wire rope. Drilling line runs from ⅞ to 2 inches (22 to 51 millimetres) in diameter and is similar to common fiber rope, but wire rope, as the name implies, is made out of steel wires (fig. 91). It looks very much like what the rest of the world calls "cable" but is designed especially for the heavy loads encountered on the rig.

The line comes off a large reel—a *supply reel* (fig. 92). From the supply reel, it goes to a strong clamp called the "deadline anchor" (fig. 93). From the deadline anchor, the drilling line runs up to the top of the mast or derrick to a set of large pulleys. This large set of pulleys is called the "crown block" (fig. 94).

Figure 92. Drilling line is stored on this supply reel at the rig. When needed, the line can be taken off the reel to replace worn line.

Figure 93. Drilling line is firmly clamped to this deadline anchor.

Figure 94. The sheaves (pulleys) of this crown block are near the bottom of the photo.

In the oilfield, the pulleys are termed "sheaves" (pronounced "shivs"). The drilling line is reeved (threaded) several times between the crown block and another large set of sheaves called the "traveling block." Because the line is reeved several times between the crown block sheaves and the traveling block sheaves, the effect is that of several lines. The heavier the anticipated loads on the traveling block, the more times the line is reeved between the crown and traveling block. For example, a deep hole, where the load on the hoisting system will be great, calls for more lines to be strung than for a shallow hole, where the load will be lighter. In figure 95, ten lines are strung, which means the line was reeved five times between the crown and traveling block. Ten lines can lift fairly heavy loads; so, only eight might be used for lighter loads. For heavier loads, twelve or more could be strung.

Figure 95. Ten lines are strung between the traveling block and the crown block.

Figure 96. Several wraps of drilling line on the drawworks drum

Once the last line has been strung over the crown block sheaves, the end of the line goes down to the drawworks drum, where it is firmly clamped. The driller then takes several wraps of line around the drum (fig. 96). The part of the drilling line running from the drawworks to the crown block is the *fastline*—fast because it moves as the driller raises or lowers the traveling block in the mast or derrick. The end of the line that runs from the crown block to the deadline anchor is the *deadline*—dead because it does not move.

If, on a rig visit, you look at the crown and traveling blocks, you may not realize just how large they are because of the distance you see them from. The sheaves around which the drilling line passes are often 5 feet (1.5 metres) or more in diameter, and the pins on which the sheaves rotate may be 1 foot (30 centimetres) or so in diameter. That's big. Incidentally, the number of sheaves on the crown block always numbers one more than the number of sheaves on the traveling block. For instance, a ten-line string requires six sheaves in the crown block and five in the traveling block. The extra sheave in the crown is needed for reeving the deadline.

Attachments to the traveling block include a spring to act as a shock absorber and a large hook from which crew members suspend the drill string (fig. 97).

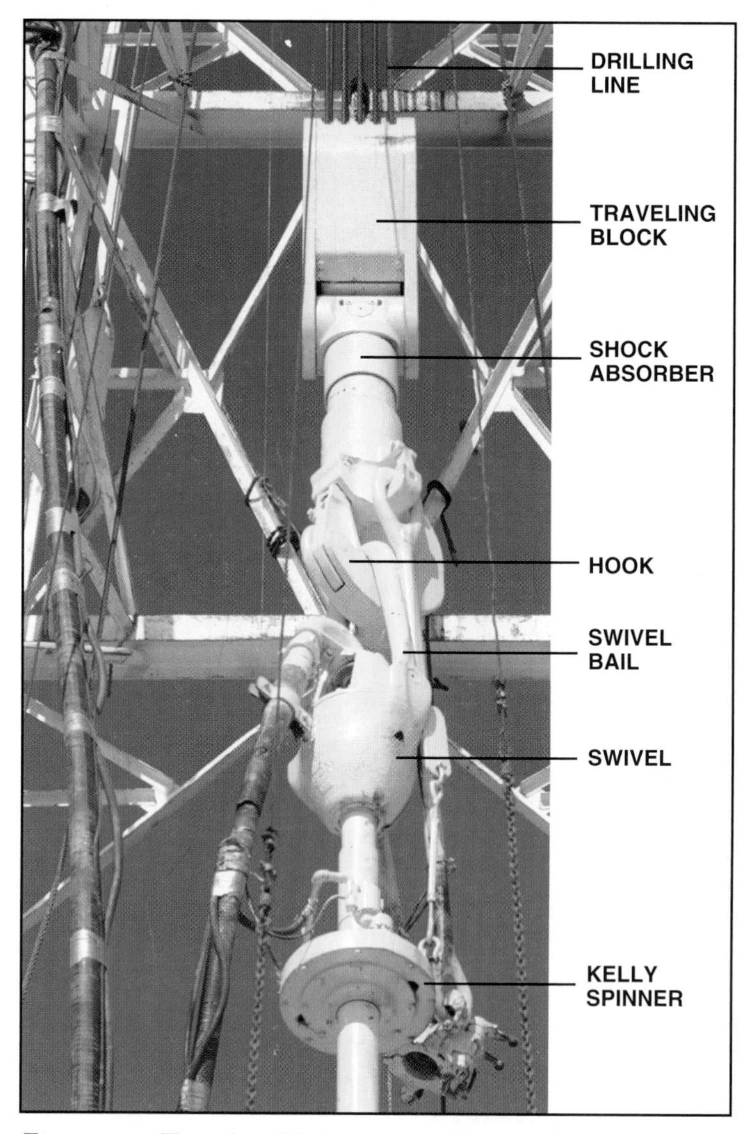

Figure 97. Traveling block, shock absorber, hook, and other equipment are suspended by wire rope drilling line.

Masts and Derricks

Masts and derricks are tall structural towers that support the blocks and drilling tools (fig. 98). They also provide height to allow the driller to raise the drill string so crew members can break it out and make it up. As mentioned earlier, a *mast* is a portable derrick that crew members can raise and lower as a unit. A *standard derrick*, on the other hand, requires that crew members assemble and disassemble it piece by piece; they cannot erect it or take it down as a single unit. Most rigs today use masts because they rig up and down much quicker than standard derricks.

Figure 98. The mast supports the blocks and other drilling tools.

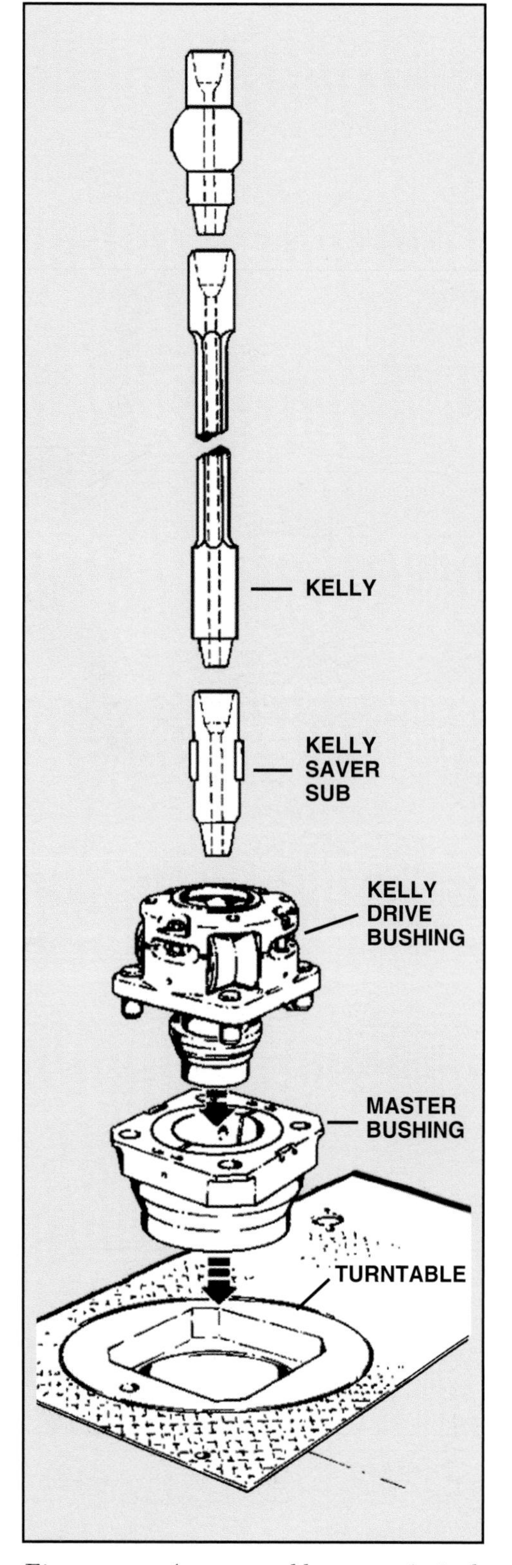

Figure 99. A rotary-table system (swivel not shown)

Manufacturers have to make masts strong and, at the same time, they have to make them portable. Consider that on a deep well, the loads that a mast may support can be as high as 2 million pounds, or 1,000 tons (907 tonnes). Yet, after finishing up one hole, crew members usually move the rig several miles to begin another. Some of the specifications manufacturers use to rate masts and derricks are their height, the vertical load they can carry, and the wind load they can withstand from the side. For example, a mast may be 136 feet (41.5 metres) tall, be able to support 275 tons (249 tonnes), and be capable of withstanding 100-mile-per-hour (161-kilometre-per-hour) winds. These specifications are impressive when you consider that such a strong piece of equipment is also relatively easy to move.

ROTATING SYSTEMS

Rotating equipment turns the bit. Generally, rigs can rotate the bit in one of three ways. The traditional way, the method that still dominates drilling, especially on land sites, uses a *rotary table* and *kelly*. A second way uses a *top-drive system*, which drilling contractors began to employ widely in the 1980s. A third way uses a *downhole motor*, which contractors use in special cases.

Rotary-Table System

Today, many contractors, especially those offshore, employ top drives on their rigs. However, many rigs still use the *rotary-table system* to rotate the drill string and bit. A rotary-table system consists of five main parts: (1) a rotary table with a turntable, (2) a master bushing, (3) a kelly drive bushing, (4) a kelly, and (5) a swivel (fig. 99).

Turntable

A stationary heavy-duty rectangular steel case houses the rotating *turntable* (fig. 100). The turntable is round in shape and is near the middle of the case. The turntable produces a turning motion that machinery transfers to the pipe and bit. The case also holds gears, bearings, and other components on which the turntable rotates. An electric motor or gears and chains from the rig drawworks power the turntable. Additional equipment transfers the turntable's turning motion to the drill pipe and attached bit.

Figure 100. The turntable is housed in a steel case.

Master Bushing and Kelly Drive Bushing

A *bushing* is a fitting that goes inside an opening in a machine. A rotary table *master bushing* fits inside the turntable (fig. 101). The turntable rotates the master bushing. The master bushing has an opening through which crew members run pipe into the wellbore.

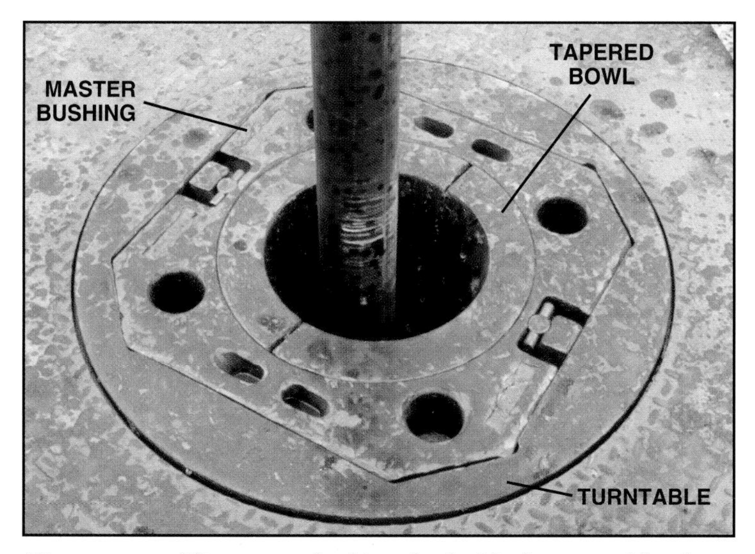

Figure 101. The master bushing fits inside the turntable. A tapered bowl fits in the master bushing.

Figure 102. Crew members are installing one of two halves that make up the tapered bowl.

A *tapered bowl* fits inside the master bushing (fig. 102). This bowl serves a vital function when the pipe and bit are not rotating. When the driller stops the rotary table and uses the rig's hoisting system to lift the pipe and bit off the bottom of the hole, it is often necessary for crew members to suspend the pipe off bottom. To do so, they place a set of segmented pipe gripping elements called "slips" around the pipe and into the master bushing's tapered bowl (fig. 103). The slips firmly grip the pipe to keep it suspended off the bottom.

The third piece of rotary equipment is the *kelly drive bushing*. A kelly drive bushing transfers the master bushing's rotation to a special length of pipe called the "kelly." The kelly drive bushing fits into the master bushing. Two types of master and kelly drive bushing are available. One master bushing has four drive holes (fig. 104). Strong steel pins on the bottom of a kelly drive bushing made for this type of master bushing fit into the holes. When the master bushing rotates, the pins engaged in the drive holes rotate the kelly drive bushing.

Figure 103. Crew members set slips around the drill pipe and inside the master bushing's tapered bowl to suspend the pipe.

Figure 104. This master bushing has four drive holes into which steel pins on the kelly drive bushing fit.

Another type of master bushing has a square opening and no drive holes. The opening corresponds to a square shape on the bottom of a kelly drive bushing made for this kind of master bushing. A square kelly drive bushing does not have drive pins. Instead, the square bottom of the kelly drive bushing fits into the corresponding square opening in the master bushing (fig. 105). With the square drive bushing in place, the rotating master bushing turns it.

Figure 105. This master bushing has a square bottom that fits into a square opening in the master bushing.

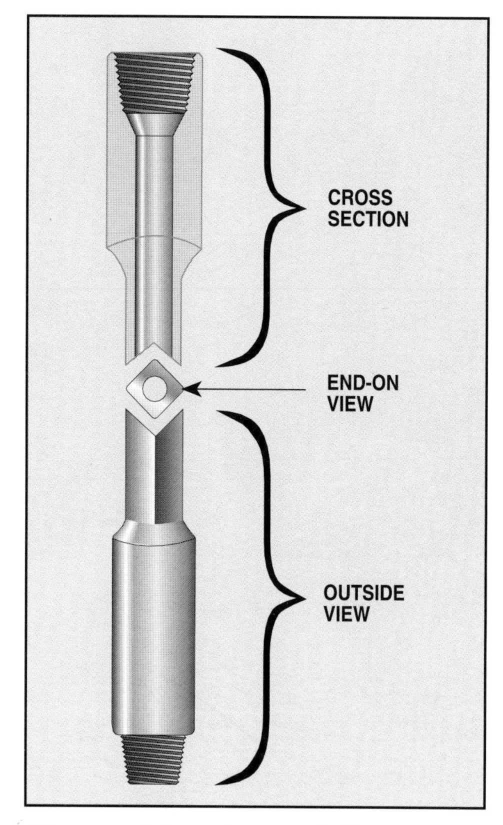

Figure 106A. A square kelly

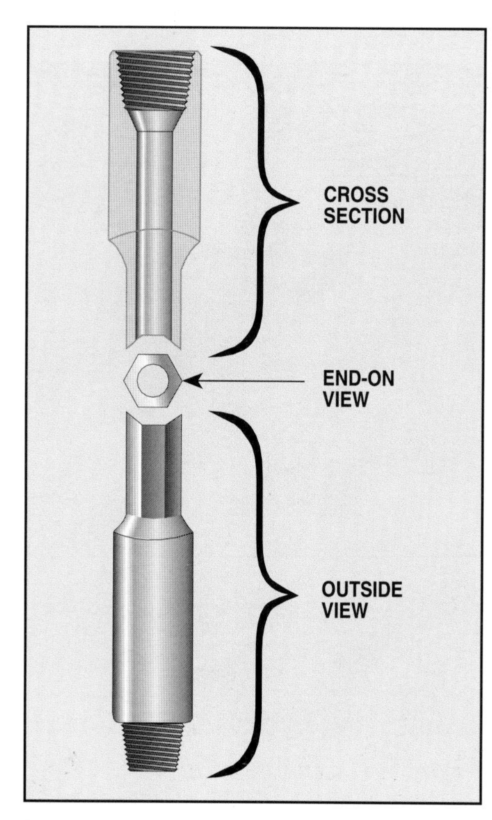

Figure 106B. A hexagonal kelly

Kelly

The fourth major part of a rotary-table system is the *kelly*. A kelly is a special length of pipe. It is not round like conventional pipe, however. Instead, it has four or six flattened sides that run almost its entire length (fig. 106 A, B). Kellys are square or hexagonal in cross section, instead of round, because the flat sides enable the kelly to be rotated. The kelly's flat sides mate with a corresponding square or hexagonal opening in the kelly drive bushing. Therefore, when the driller inserts a square or hexagonal kelly into the matching four- or six-sided opening in the kelly drive bushing and activates the rotary table, the kelly drive bushing rotates the kelly (fig. 107). Moreover, because crew members make up the drill string to the kelly, the kelly rotates the string and attached bit.

The kelly slides easily into the drive-bushing opening. It is therefore free to move up or down through the bushing opening, even as it rotates. The kelly's being able to move through the rotating bushing is important because it allows the kelly to follow the bit down as it drills deeper. When the driller stops

Figure 107. This hexagonal kelly fits inside a matching opening in top of the kelly drive bushing. Thus, when the bushing rotates, so does the kelly.

rotating and lifts (*picks up*) the drill string, the kelly drive bushing slides along the kelly as the hoisting system raises the kelly. When the drive bushing reaches the bottom of the kelly, the kelly tool joint, being bigger than the bushing's opening, keeps the bushing from sliding off the kelly.

In general, a hexagonal kelly is stronger than a square kelly. Consequently, contractors tend to use hexagonal kellys on large rigs to drill deep wells because of their extra strength. Small rigs often use square kellys because they are less expensive. Manufacturers make most square and hexagonal kellys to American Petroleum Institute (API) specifications. (The API is a trade association that sets oilfield standards and specifications. A standard API kelly, either square or hexagonal, is 40 feet (12.2 metres) long, although an optional length of 54 feet (16.5 metres) is also available. Most rigs use 40-foot kellys.

Swivel

The fifth principal part of a rotary-table system is the *swivel.* The swivel interfaces the rotary system with the hoisting system. A heavy-duty *bail*, similar to the bail, or handle, on a water bucket but much larger, fits into a big *hook* on the bottom of the traveling block (fig. 108). The hook suspends the swivel and attached drill string.

Crew members make up the top of the kelly to the swivel. The kelly screws onto a threaded fitting—the *stem*—that comes out of the swivel. This stem rotates with the kelly, the drill string, and the bit. At the same time, drilling mud flows through the stem and into the kelly and drill string.

Figure 108. The hook on the bottom of the traveling block is about to be latched onto the bail of the swivel.

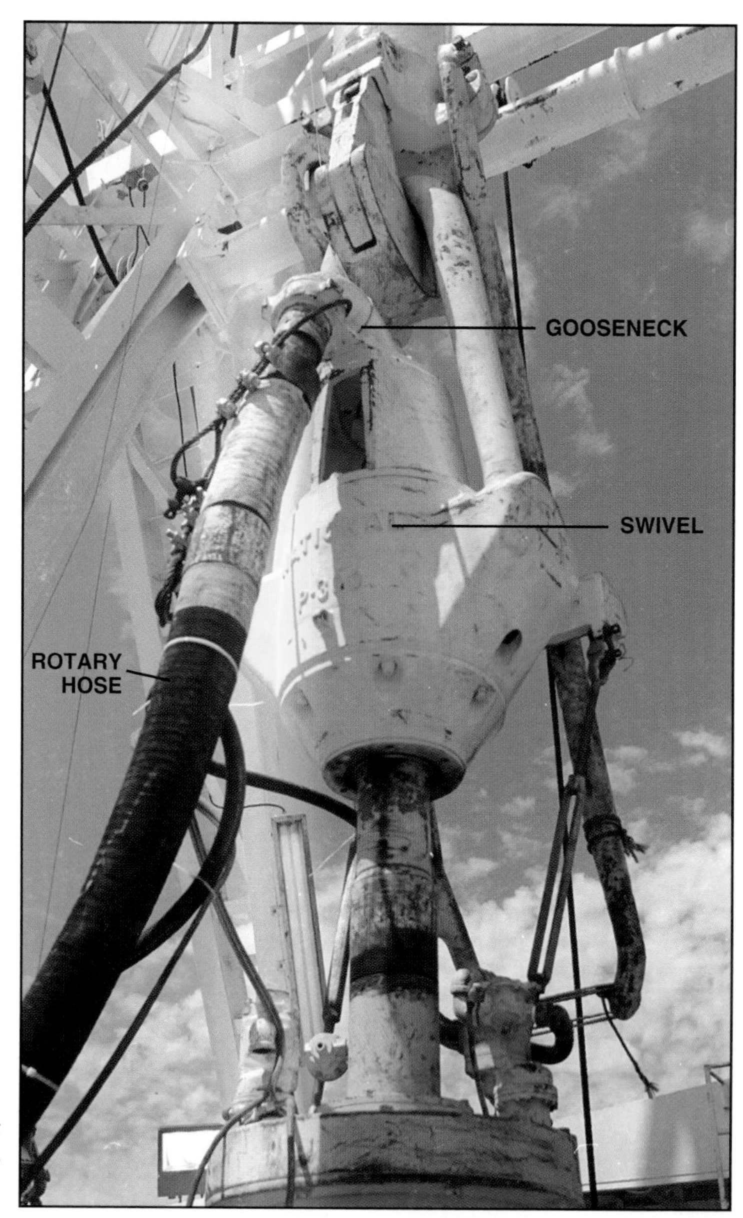

Figure 109. Drilling fluid goes through the rotary hose and enters the swivel through the gooseneck.

Near the top and on one side of the swivel is a *gooseneck* (fig. 109). The gooseneck is a curved, erosion-resistant piece of pipe. It conducts drilling mud under high pressure into the swivel stem. A special hose—the *rotary*, or *kelly*, *hose*—attaches to the gooseneck. The rotary hose conducts drilling mud from the pump to the swivel.

To summarize the kelly-and-rotary-table system: (a) the turntable in the rotary table rotates the master bushing; (b) the master bushing rotates the kelly drive bushing; (c) the kelly drive bushing rotates the kelly; (d) the kelly rotates the attached pipe and bit; and (e) the swivel suspends the pipe, allows it to rotate, and has a passage for drilling mud to enter the kelly and pipe.

Top Drives

While many rigs use the kelly-and-rotary-table system to rotate the drill string and bit, a large number use a different system. This system does away with the kelly and thus the kelly drive bushing and a rotating master bushing. Instead, a *top drive*, also called a "power swivel," rotates the drill string and bit (fig. 110). Like a regular swivel, a top drive hangs from the rig's large hook and it has a passageway for drilling mud to get into the drill pipe. However, a top drive comes equipped with a heavy-duty electric motor (some large top drives have two motors). Drillers operate the top drive from their control console on the rig floor.

The motor turns a threaded drive shaft. The crew *stabs* (inserts) the drive shaft into the top of the drill string. When the driller starts the top drive's motor, it rotates the drill string and the bit. A top drive eliminates the need for a conventional swivel, a kelly, a rotating master bushing, and a kelly drive bushing.

Figure 110. A top drive, or power swivel, hangs from the traveling block and hook.

Rigs with a top drive, however, still need a rotary table with a master bushing and bowl to provide a place to suspend the pipe on slips when the bit is not drilling. Because the rotary table only serves as a place for crew members to place slips on rigs with top drives, manufacturers make special *rotary support tables* for top-drive rigs that are lighter and smaller than regular rotary tables. They are still rugged enough, however, to support the weight of the drill string. Further, some have built-in hydraulic motors that can rotate the turntable should the top-drive system malfunction. Hydraulic motors are considerably lighter in weight than electric motors and take up less space.

The main advantage of a top drive over a kelly-and-rotary-table system is that a top drive makes it safer and easier for crew members to handle the pipe. Because of the way in which a rig with a rotary-table system operates, the crew can add only one joint of drill pipe at a time as the hole deepens. With a top-drive system, on the other hand, because it operates differently from the conventional system, the crew can add pipe three joints at a time, if they choose to do so. Adding three-joint stands of pipe saves time.

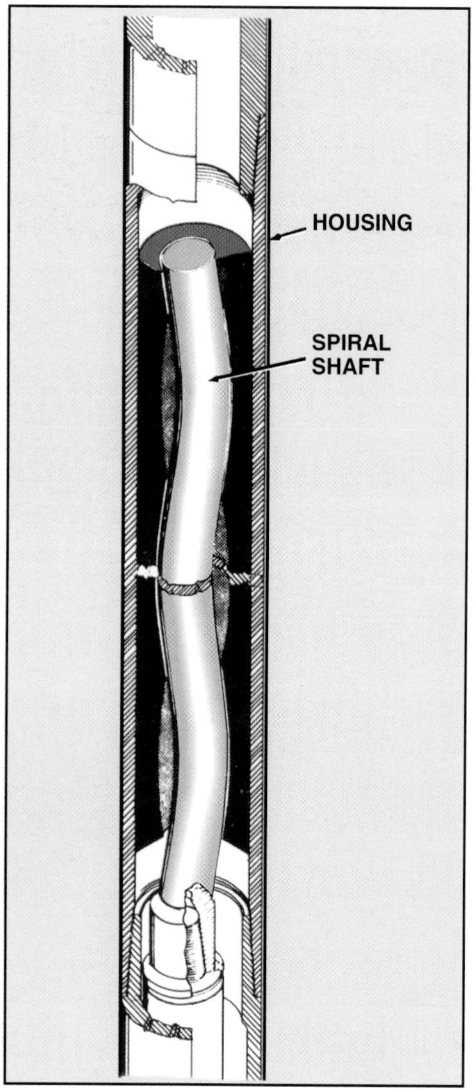

Figure 111. Mud pressure pumped through the drill stem above the downhole motor, forces the motor's spiral shaft to turn inside the housing. The turning shaft turns the bit (not shown), which is attached to the bottom of the shaft.

Downhole Motors

In special situations, the rig may use a *downhole motor* to rotate the bit. Unlike a rotary-table or a top-drive system, a downhole motor does not rotate the drill pipe. Instead, it rotates only the bit. Drilling mud powers most downhole motors. Normally, crew members install the motor in the drill string just above the bit.

To make a mud motor rotate the bit, the driller pumps drilling mud down the drill string as usual. When the mud enters the motor, however, it strikes a spiral shaft, which goes inside a tubular housing (fig. 111). The shaft and housing fit in such a way that mud pressure causes the shaft to turn. Because the bit is attached to the motor shaft, the shaft turns the bit. The mud exits out of the bit as usual.

Rigs often use downhole motors to drill directional holes. A directional hole is a hole that is intentionally drilled off-vertical. Sometimes, it is desirable to drill a hole on a slant because a vertical hole could not reach a desired part of a petroleum reservoir. Because it is easier to get the bit to drill in the desired direction if the drill string does not rotate, rigs employ downhole motors. One major instance of directional drilling is drilling horizontal holes. A horizontal hole drilled through a formation can, under the right conditions, allow a company to produce the formation much better than with a vertical hole (fig. 112).

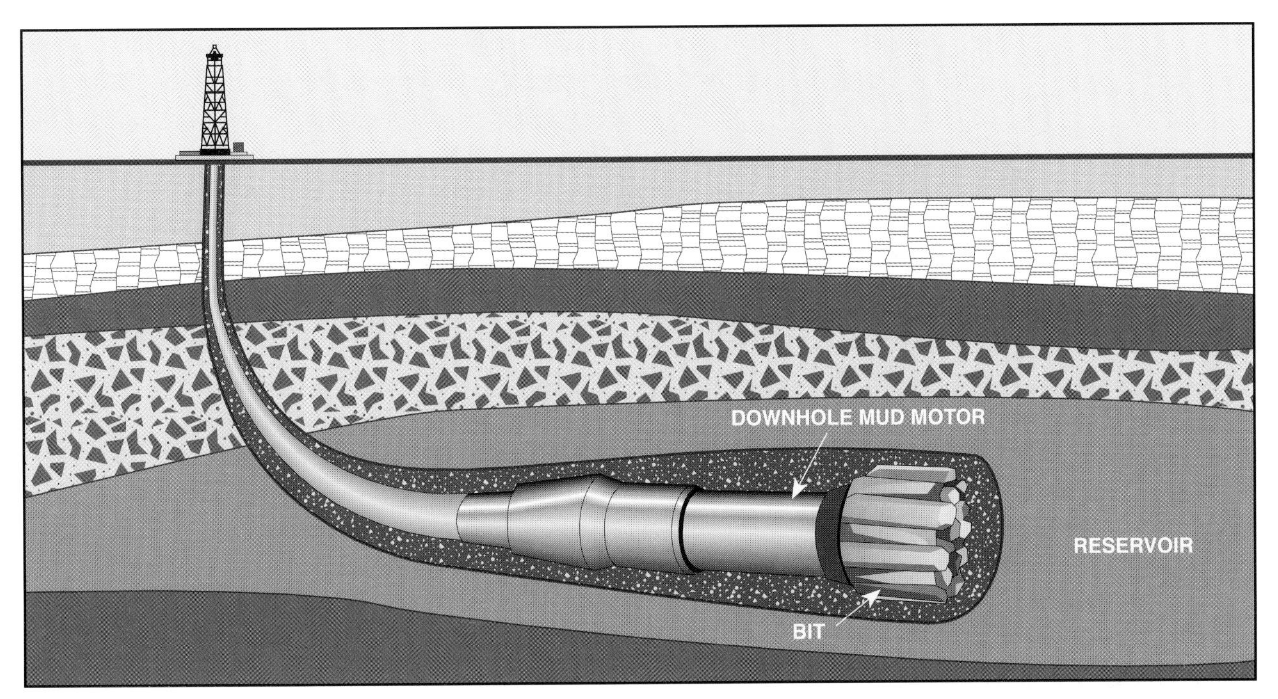

Figure 112. Horizontal hole

The Drill String

The drill string consists of *drill pipe* and special, heavy-walled pipe called "drill collars" (fig. 113). Manufacturers make most drill pipe from steel, but they also use aluminum. Drill collars, like drill pipe, are metal tubes through which the driller pumps drilling fluid. They are heavier than drill pipe, however.

Figure 113. These drill collars are laid out on a rack prior to being run into the well.

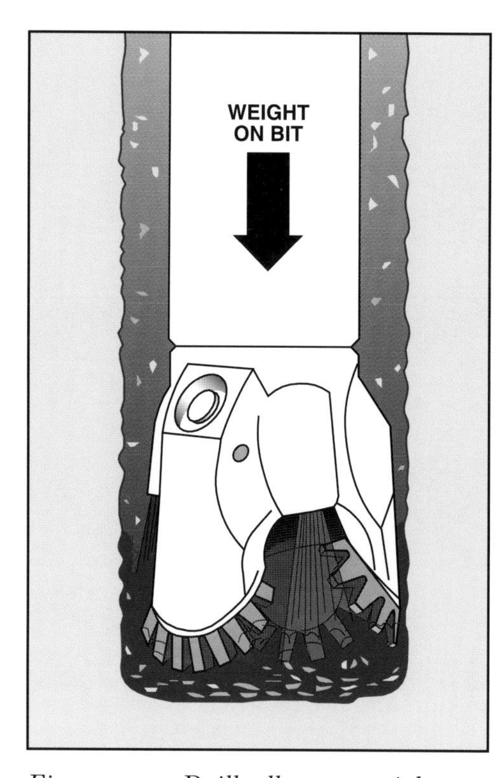

Figure 114. Drill collars put weight on the bit, which forces the bit cutters into the formation to drill it.

Drill collars are heavy because they are used in the bottom part of the string to put weight on the bit. This weight presses down on the bit so the bit cutters can bite into the formation and drill it (fig. 114). Most of the drill string is made up of drill pipe but crew members make up enough drill collars to put the required weight on the bit.

Drill collars are either 30 or 31 feet (9.1 or 9.4 metres) long and those made to API specifications range in diameter from 2⅞ to 12 inches (73.03 to 304.8 millimetres). To give you an idea of a drill collar's weight, one that is 30 feet (9.1 metres) long and 6 inches (152.4 millimetres) in diameter weighs about 3,000 pounds (1,361 kilograms). Thus, if the drill crew made up ten joints of this particular drill collar, the assembly would weigh 30,000 pounds (13, 610 kilograms). The amount of weight a bit requires to drill efficiently varies considerably and depends on the type of bit and the type of formation it is drilling. Nevertheless, 30,000 pounds is a good example of bit weight required.

A length of drill pipe is about 30 feet (9.1 metres) long, and drilling people call each length a "joint of pipe" (fig. 115). Each end of each joint is threaded. One end has inside or female threads; the other has outside or male threads. The

Figure 115. Several joints of drill pipe are laid on the rack prior to being run into the hole.

female end is called the "box," and the male end is called the "pin." When crew members make up drill pipe, they insert, or *stab*, the pin end in the box and tighten the connection (fig. 116). Crew members call the threaded ends of drill pipe "tool joints." Normally, the manufacturer welds the tool joints onto the ends of the drill pipe and cuts the threads to API specifications.

Figure 116. A floorhand stabs the pin of a joint of drill pipe into the box of another joint.

Figure 117. Two drill collars on a pipe rack; at left is the drill collar box; at right is the pin.

Manufacturers do not add tool joints to drill collars. The walls of drill collars are so thick that it is not necessary. Instead, the manufacturer cuts the threads directly onto and in the drill collars. Like drill pipe, drill collars also have a box and pin end (fig. 117). Thus, you can easily distinguish drill pipe from drill collars because drill collars do not have the bulge at either end that characterizes the tool joints of drill pipe (fig. 118).

Figure 118. Drill collars stacked in front of drill pipe on the rig floor

Bits

A rig's primary job is to rotate a bit on the bottom of the hole. The bit is the business end of a drilling rig, because the bit drills, or makes, the hole. Bit manufacturers offer several types of bit in many sizes. They design them to drill a particular size of hole in a particular kind of formation.

Bits fall into two main categories: (1) roller cone and (2) fixed head. Both have *cutters*, which remove rock as the bit drills. Bits have several kinds of cutters. Cutters for roller cone bits are either steel teeth or tungsten carbide inserts. Cutters for fixed-head bits are natural diamonds, synthetic diamonds, or a combination (a *hybrid*) of cutters. Hybrid bits combine natural and synthetic diamonds, and may have tungsten carbide inserts. Table 3 lists bits and cutters.

Table 3
Roller cone and fixed-head bit cutters

Roller Cone Bits	Fixed-Head Bits
Steel teeth	Natural diamonds
Tungsten carbide inserts	Synthetic diamonds
	Hybrid

Roller Cone Bits

Roller cone bits have *steel cones* that roll, or turn, as the bit rotates (fig. 119). The bit cutters are on the cones. As the cones roll over the bottom of the hole, the cutters scrape, gouge, or crush the rock into relatively small chips, or *cuttings*. Drilling fluid, which comes out of special openings in the bit, removes the cuttings. Most roller cone bits have three cones, although some have two or four. The cutters on a roller cone bit are either steel teeth or tungsten carbide inserts. Manufacturers *mill* (cut) or *forge* (hammer) the teeth out of the steel body of the cones. For tungsten carbide insert bits, they drill holes in the cones and press-fit the tungsten carbide cutters into the holes (fig. 120).

Figure 119. A roller cone bit has teeth (cutters) that roll, or turn, as the bit rotates.

Figure 120. Tungsten carbide inserts are tightly pressed into holes drilled into the bit cones.

Figure 121. Drilling fluid (salt water in this case) is ejected out of the nozzles of a roller cone bit.

Most steel-tooth and tungsten carbide roller cone bits have nozzles that eject high-speed streams, or *jets*, of drilling mud (fig. 121). The jets of mud sweep cuttings out of the way as the bit drills. With the cuttings out of the way, the bit cutters do not redrill cuttings, which would slow the drilling rate, or *rate of penetration* (*ROP*—pronounce each letter). Because of their jetting action, oil people sometimes call roller cone bits "jet bits."

Fixed-Head Bits

Although fixed-head bits have jets, they do not have cones that roll independently on the bit as it rotates. Instead, fixed-head bits consist of a solid piece—the *head*—that rotates only as the drill string rotates. The bit manufacturer sets the cutters into the bit head. One type of fixed-head bit has natural, industrial-grade diamond cutters. Another employs synthetic diamonds. One synthetic diamond is *polycrystalline diamond*; manufacturers also use a synthetic diamond called a "thermally stable polycrystalline diamond."

In a natural diamond bit, the bit maker embeds industrial diamonds in the bottom and sides of the bit head. As the bit rotates, the diamonds contact the face of the formation and plow and grind it to make hole. Manufacturers make many kinds of diamond bits for many kinds of formations and drilling conditions (fig. 122).

A widely used bit is the *polycrystalline diamond compact* (*PDC*) bit. PDC bits feature tungsten carbide *compacts* to which are bonded synthetic diamonds. (A compact, in this case, is a small disk made of tungsten carbide.) The manufacturer machines

Figure 122. Several types of diamond bit are available.

pockets into ribs or blades on the bit body and inserts the diamond-coated compacts into the pockets (fig. 123). A special kind of PDC bit is a *thermally stable polycrystalline diamond* (*TSP*) bit. TSP bits can withstand much higher temperatures than PDC bits. Thus, when drilling a hole that requires a lot of weight and high rotary speeds that generate enough heat to destroy the synthetic diamond coating of a PDC cutter, the operator may select a TSP bit.

Bit designers have taken advantage of the unique properties of each of the various materials used to make fixed-head bit cutters and have combined them on one bit. Called "hybrid bits," they combine natural diamonds, PDCs, TSPs, and even tungsten carbide inserts.

Operators use natural diamond, PDC, TSP, and hybrid bits to drill soft, medium, and hard formations. They are especially effective in abrasive formations. These natural and synthetic diamond bits are the most expensive type of bit. When properly used, however, they can often drill faster and last longer than steel-tooth or carbide insert bits.

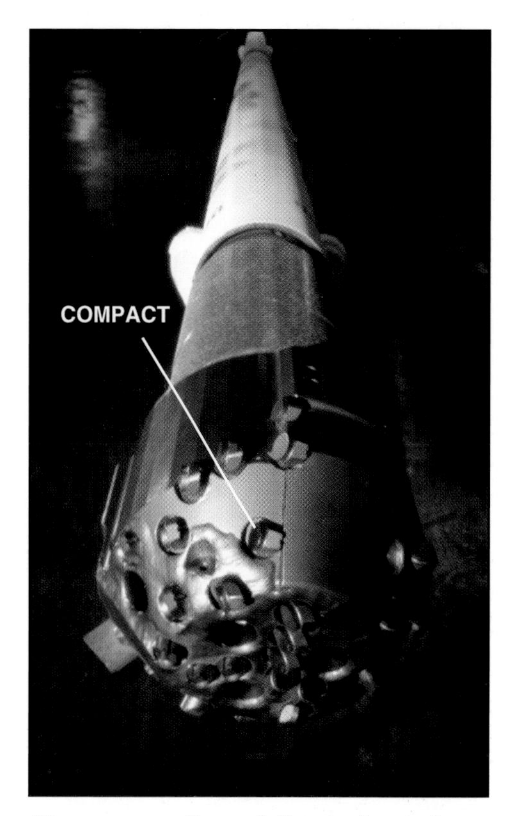

Figure 123. Several diamond coated tungsten carbide disks (compacts) form the cutters on this polycrystalline diamond compact (PDC) bit.

Bit Sizes and Attributes

Bits are available in many sizes, from 3¾ inches (95.25 millimetres) to 28 inches (711.2 millimetres) in diameter, depending on the diameter of the hole the operator needs to have drilled. An operator or contractor can special order smaller or larger sizes if required. Moreover, because formations of various hardnesses exist, manufacturers also offer bits with cutters designed to drill formations of different hardnesses. In general, they offer bits with cutters suited to drill soft, medium soft, medium, medium hard, hard, and very hard and abrasive formations.

Weight on Bit and Rotating Speeds

Putting weight on a bit makes its cutters bite into the rock. Usually, drillers apply weight on the bit by allowing some of the weight of the drill collars above the bit to press down on it. The amount of weight depends on the size and the type of bit and the speed at which the driller rotates it. The amount of weight also depends on the type of formation being drilled. For a general idea, consider that a driller can rotate bits anywhere from 70 to 180 revolutions per minute and can place anywhere from 26,000 to 130,000 pounds (11,570 to 57,850 decanewtons) of weight or force on them.

CIRCULATING SYSTEM

One unique characteristic of rotary drilling is the pumping of drilling fluid to the bottom of the hole to pick up cuttings made by the bit and lift them to the surface for disposal. At the same time, solid particles in the fluid plaster the sides of the hole opposite porous and permeable formations and keep them from caving in. The ability of a rotary rig to circulate drilling fluid has made it the drilling method of choice all over the world.

Drilling Fluid

Drilling fluid—*mud*—is usually a mixture of water, clay, weighting material, and a few chemicals. During drilling it circulates in steel tanks (fig. 124). Some formations swell in the presence of water and impede drilling, so the operating company requires that the contractor use oil instead of water as a base for the mud. In a few cases, the operator may choose air or gas as a drilling fluid. (A fluid can be either a gas or a liquid. Air is a gaseous fluid; water is a liquid fluid.) Unlike mud, air or gas exerts very little pressure on the bottom of the hole because it is so much lighter in weight (it is less dense) than mud. Accordingly, air or gas can dramatically increase the drilling rate, or rate of penetration. Air or gas, because they are so light in density, allow bit cuttings to move rapidly away from the bit. With few cuttings getting in the way of the bit cutters, the cutters always contact fresh, uncut formation.

With such an advantage, you would think that companies would drill every hole with air or gas. Sadly, drilling with air or gas as a circulating fluid has a major drawback. Water in a formation can enter the hole, wet the cuttings, and cause them to ball up. If enough water enters, the balled up cuttings clog

Figure 124. Drilling mud swirls in one of several steel tanks on this rig.

the hole and prevent circulation. Without circulation, drilling stops. Unfortunately, most subsurface formations contain ample amounts of water; consequently, air and gas drilling has limited use.

Whether gaseous or liquid, drilling fluid plays several vital roles in rotary drilling. It raises cuttings made by the bit to the surface. It also cools and lubricates the rotating drill stem and bit. Moreover, drilling mud keeps underground pressure in check. A hole full of drilling mud exerts pressure, just as a swimming pool full of water exerts pressure. Mud pressure in the borehole offsets the pressure in a formation.

The heavier, or denser, a mud is, the more pressure it exerts. Water or oil by itself often does not weigh enough to exert the necessary pressure, especially as the hole gets deep. A gallon of fresh water, for instance, only weighs about 8⅓ pounds. (A cubic metre of water weighs about 1,000 kilograms.) To make water or oil exert the correct amount of pressure—not too little and not too much—the operator has the derrickman add weighting material. A mineral called "barite" is a popular weighting material. It is over four times heavier than water. Barite is supplied to the rig as a fine powder, and the derrickman gradually adds it to the mud's water or oil. The mud suspends the powdered barite uniformly throughout the hole.

At the rig, you may hear personnel talk about "mud weight," which is another way of saying mud density. As mentioned earlier, mud weight is important because it indicates how much pressure the mud exerts to hold formation pressure in check. In most of the U.S., oilfield hands measure mud weight in pounds per gallon (fig. 125). California is different, of course; there, they measure it in pounds per cubic foot.

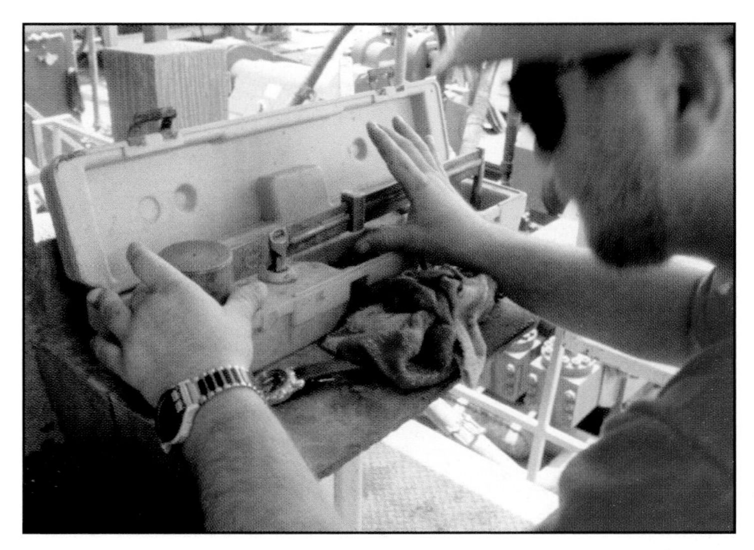

Figure 125. This derrickman is measuring the density (weight) of a sample of drilling mud in a balance that is calibrated in pounds per gallon.

In Canada, and many other countries, they measure it in kilograms per cubic metre. So a mud that weighs 10 pounds per gallon also weighs 74.8 pounds per cubic foot or 1,198.2 kilograms per cubic metre.

With instructions from the operator and possibly a drilling fluids engineer, the derrickman also adds special *clay* to the water or oil. This clay, when thoroughly mixed into the mud, keeps the cuttings in suspension as they move up the hole. When the driller stops pumping the mud for any reason, the clay makes the mud temporarily *gel*. The gelled mud keeps the cuttings suspended even when the mud is not moving. When the driller *breaks circulation* (starts pumping again), the mud liquefies, or ungels, to move up the hole.

As mentioned before, particles of clay also line the wall of the hole, much as plaster sticks to the wall of a room. The clay solids form a thin but strong lining, or wall cake, that stabilizes the hole and keeps it from caving in, or *sloughing*. Wall cake may be only a few thirty-seconds of an inch (a few millimetres) thick, but the role it plays in stabilizing the hole is vital.

Circulating Equipment

Mud circulates through many pieces of equipment, all of which play an important role. Circulating equipment includes the mud pump, the discharge line, the standpipe, the rotary hose, the swivel (or top drive), and the kelly (on rigs with a rotary- table system), the drill pipe, the drill collars, the bit, the annulus, the return line, the shale shaker, the desilter, the desander, the mud tanks, and the suction line (fig. 126). The *mud pump* (fig. 127) takes mud from the mud tanks and sends it out a discharge line to a *standpipe*. The standpipe is a steel pipe mounted vertically on one leg of the mast or derrick (fig. 128). Mud flows out of the standpipe and into the rotary hose, which is connected to the swivel on rotary-table system rigs or to the top drive. Mud goes down the kelly on rigs with a rotary table; on rigs with a top drive, mud goes through passageways inside it. Once it leaves the kelly or the top drive, mud flows down the drill stem, out the bit. It does a sharp U-turn and heads back up the hole in the *annulus*. The annulus is the space between the outside of the drill string and sides of the hole. As it flows up the annulus, the mud carries the cuttings made by the bit.

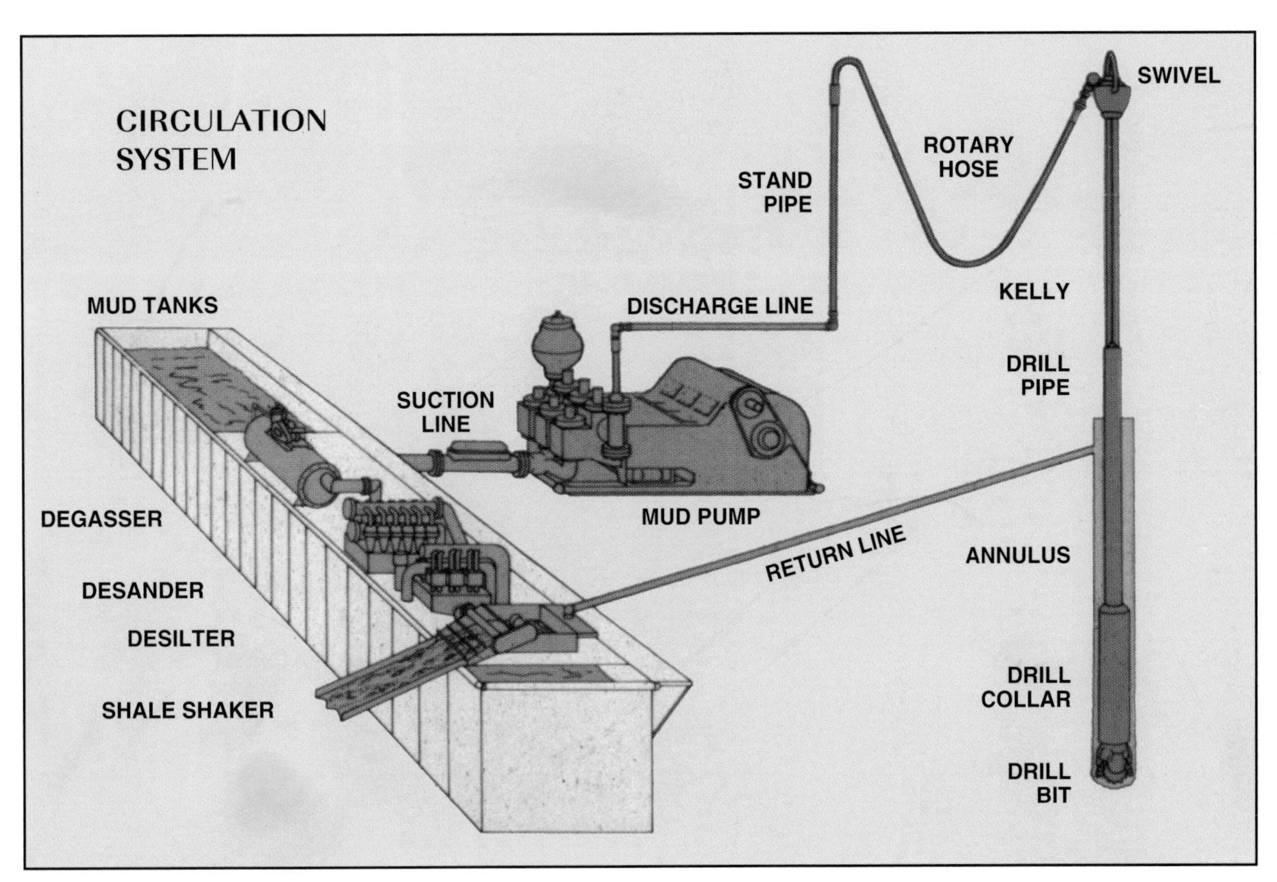

Figure 126. Components of a rig circulating system

Figure 127. Mud pumps (most rigs have at least two) are powerful machines that move drilling mud through the circulating system.

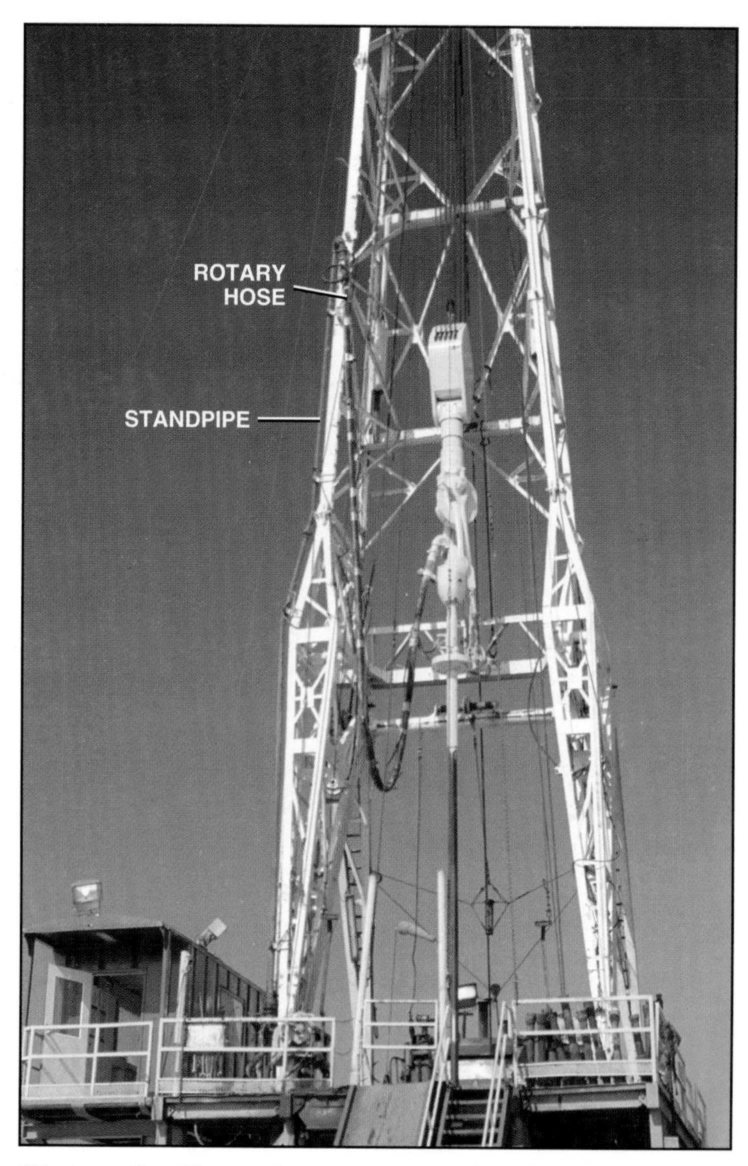

Figure 128. The standpipe runs up one leg of the derrick or mast and conducts mud from the pump to the rotary hose.

Finally, the mud leaves the hole through a steel pipe called the "mud return line" and falls over a vibrating, screenlike device called the "shale shaker" (fig. 129). The shale shaker is appropriately named, for it rapidly vibrates or shakes as the mud returning from the hole falls over it. The shale shaker acts like a sifter and screens out the cuttings. Except in environmentally sensitive areas on land, the cuttings fall into the reserve pit, the earthen pit excavated when the site was being prepared. In areas where the contractor cannot use a reserve pit because of environmental reasons, the shaker dumps the cuttings into a special receptacle.

Figure 129. Mud and cuttings fall onto the shale shaker, which removes the cuttings. The mud falls into a tank below the shaker.

Later, the cuttings are properly disposed of. Offshore, the cuttings are usually dumped into a barge to be transported to a land site for proper disposal. In either case, the mud drains back into the mud tanks where the mud pump recycles it downhole. The circulating system is essentially a closed system. The system circulates the mud over and over throughout the drilling of the well. From time to time, however, crew members may add water, clay, or other chemicals to make up for losses or to adjust the mud's properties as the hole drills into new and different formations.

Figure 130. Desanders remove fine particles, or solids, from drilling mud.

Figure 131. Desilters, like desanders, also remove fine solids from the mud; desilters remove solids that are even smaller than those the desander removes.

Auxiliary Equipment

Several pieces of auxiliary equipment keep the mud in good shape. The shale shaker sifts out the normal-sized cuttings. Sometimes, though, the bit creates particles so small that they fall through the shaker with the mud. So, after the mud passes through the shale shaker, the system sends the mud through *desanders* (fig. 130), *desilters* (fig. 131), *mud cleaners* (fig. 132), and *mud centrifuges* (fig. 133). These pieces of equipment remove fine particles, or small solids, to keep them from contaminating the drilling mud.

Figure 132. Mud cleaners, like desanders and desilters, also remove small solid particles from the mud.

Figure 133. A mud centrifuge removes tiny solids from the mud; contractors also use centrifuges to recover weighting material (barite) from the mud.

A *degasser* (fig. 134) removes small amounts of gas that enter the drilling mud as it circulates past a formation that contains gas. A degasser is used when the amount of gas is not enough to make the well a producer; instead, it is just enough to contaminate the mud. The driller cannot recirculate this gas-cut mud back into the hole because the gas makes the mud lighter, or less dense. If the mud gets too light, the well can *kick*—formation fluids under pressure can enter the hole. If not handled properly, a kick can lead to a *blowout*. (The Spindletop well was a blowout. Oil gushed out of the well at an uncontrolled rate and did so until Lucas and his workers could *cap* it—that is, put a heavy-duty valve on top of the well and close it.)

Figure 134. A degasser removes relatively small volumes of gas that enter the mud from a downhole formation and are circulated to the surface in the annulus.

Figure 135. A derrickman, wearing personal protective equipment, adds dry components to the mud through a hopper.

To add noncorrosive and noncaustic powdered components to the drilling mud, the derrickman often uses a *mud hopper* (fig. 135). The derrickman opens the sack of material, places it at the top of the hopper's large funnel, and gradually adds the powder to the funnel. At the bottom of the hopper, a high-speed stream of mud picks up the powdered material, thoroughly mixes it, and puts it into the mud tanks. Mud components that are needed in large quantities, such as clay or weighting materials, are usually stored in large tanks called "P-tanks" (fig. 136). As for caustic components, the derrickman usually adds them directly to the mud tanks through a special chemical barrel (fig. 137).

With knowledge of the power system, hoisting equipment, rotating components, and circulating equipment, the next step is to learn how crew members use these components to make hole. So, let's look next at normal drilling operations.

Figure 136. Large quantities of dry mud components are stored in P-tanks.

Figure 137. This derrickman stands next to a small, red-colored, steel container (a "barrel") through which he can add caustic materials to the mud that is in the tanks below the grating.

Normal Drilling Operations

10

A rig has a lot of equipment, and crew members have to put this equipment to work to drill a well. This section covers normal drilling operations. For our purposes, normal drilling operations include (1) drilling the hole; (2) adding a new joint of pipe as the hole deepens; (3) tripping the drill string out the hole to put on a new bit and running it back to bottom, or making a *round trip*; and (4) running and cementing casing, the large-diameter steel pipe that crew members put into the hole at various, predetermined intervals. Usually, operating companies hire a special casing crew to run the casing and they engage the services of a cementing company to place the cement around the casing. Nevertheless, the rig crew usually assists in running casing and cementing it in the well.

DRILLING THE SURFACE HOLE

To begin, assume the rig crew is ready to begin drilling the first part of the hole. In our case, let's suppose that the rathole crew prepared the initial 50 feet (15 metres) of hole. They drilled and lined it with conductor pipe as described in the section on preparing the drilling site (see fig. 66). The diameter of the conductor pipe varies, and its diameter depends on many factors, but it is usually large. In our case, let's assume it is 20 inches (508 millimetres). Therefore, the first bit the crew runs into the conductor pipe will have to be smaller than 20 inches. In this case, let's say they use a 17½-inch (444.5-millimetre) bit.

They make up this bit on the end of the first drill collar, and they lower both bit and drill collar into the conductor hole (fig. 138). They make up enough collars and drill pipe to lower the bit to bottom. On a rig using a rotary table and kelly, the driller then picks up the kelly out of the rathole where it has been waiting (fig. 139). Crew members then stab and make up the kelly onto the topmost joint of drill pipe sticking up out of the rotary table. The slips suspend this joint (and the entire drill string) in the rotary table (fig. 140).

Figure 138. A bit being lowered into the hole on a drill collar

Figure 139. The kelly and related equipment in the rathole; the kelly is not visible because it is inside the rathole that extends below the rig floor.

Figure 140. Red-painted slips with three handgrips suspend the drill string in the hole.

With the kelly made up, the driller starts the mud pump, lowers the kelly drive bushing to engage the master bushing. The driller actuates the rotary table to begin rotating the drill stem and bit (fig. 141). The driller gradually releases the drawworks brake, and the rotating bit touches bottom and begins making hole.

The sequence with a top drive is much the same as with a rotary table and kelly. The crew stabs and makes up the last joint of drill pipe onto the drive stem of the top drive (fig. 142). The driller then starts the motor in the top drive to rotate the string and bit, begins circulating mud, and lowers the assembly to bottom.

Figure 141. The kelly drive bushing is about to engage the master bushing on the rotary table; when the driller actuates the rotary table, the bit and drill string will turn.

Figure 142. The motor in the top drive turns the drill stem and bit.

Figure 143. The red pointer on the weight indicator shows weight on the bit. Here, it is a little more than 52,000 pounds. The black pointer indicates hook load, which is the amount of weight suspended by the hook. Hook load in this case is about 350,000 pounds.

With both a top drive and a rotary table system, using an instrument called the "weight indicator," the driller monitors the amount of weight put on the bit by the drill collars (fig. 143). After the bit drills about 30 feet (9 metres), which is about the length of a joint of drill pipe, crew members must add a new joint of pipe to drill deeper. On rigs with a rotary table, crew members say that the "kelly is drilled down," meaning that the bit has made enough hole so that the top of the kelly is very near the kelly drive bushing (fig. 144).

Figure 144. The kelly is drilled down (is very close to the kelly drive bushing), so it is time to make a connection.

With the kelly (or the joint of drill pipe on top-drive rigs) drilled down, the driller stops rotating, picks up (hoists) the drill string, and stops the mud pump. The floorhands are ready to *make a connection*—that is, they are ready to add (connect) a new joint of drill pipe to the drill string so that the bit can drill another 30 feet or so.

To make a connection on a rig with a rotary table and kelly, the driller picks up the drill string high enough for the kelly to clear the rotary table—that is, the driller uses the drawworks to hoist the traveling block, hook, and swivel up into the derrick or mast so that the first joint of drill pipe is exposed in the opening in the rotary table (fig. 145).

Figure 145. Here, the driller has raised the kelly with the traveling block so that the first joint of drill pipe is exposed in the opening of the rotary table.

Figure 146. Crew members latch tongs on the kelly and on the drill pipe.

With the kelly clear of the rotary table, the floorhands set the slips around the joint of drill pipe to suspend the drill string in the hole. They then latch two big wrenches called "tongs" on the kelly joint and the tool joint of the joint of drill pipe (fig. 146). A *tong pull line*—a length of strong wire rope—runs from the end of the tongs to the breakout cathead on the drawworks. The driller engages the automatic cathead, and it starts pulling on the line with tremendous force. The pulling force on the tongs breaks out (loosens) the threaded joint of the kelly and drill pipe. Once the joint is loosened, the driller engages a *kelly spinner*, which is a special air motor mounted near the top of the kelly (fig. 147). The kelly spinner rapidly turns or spins the kelly to back it out (unscrew it) from the drill pipe joint.

With the kelly backed out of the drill pipe's tool joint, crew members swing the kelly over to the mousehole, that lined hole in the rig floor the crew prepared when they rigged up. Earlier, crew members placed a joint of drill pipe into the mousehole so that it would be ready to add to the drill string.

Figure 147. The kelly spinner rapidly rotates (spins) the kelly in or out of the drill pipe joint.

They stab the kelly into the joint in the mousehole (fig. 148), and the driller spins up the kelly into the joint using the kelly spinner. Crew members grab the tongs, latch them onto the kelly and pipe, and buck up (tighten) the joint to final tightness.

Next, the driller uses the drawworks to raise the kelly and attached joint out of the mousehole. The crew stabs the end of the new joint into the joint suspended by the slips in the rotary table, and, using a spinning wrench and the tongs, they thread the joints together and buck them up to final tightness (fig. 149). Finally, the driller lifts up the kelly and attached string a small amount, the crew removes the slips (fig. 150),

Figure 148. Crew members stab the kelly into the joint of pipe in the mousehole. Note the tongs latched onto the joint.

Figure 149. Crew members use tongs to buck up (tighten) one drill pipe joint to another.

Figure 150. Crew members remove the slips.

Figure 151. The kelly drive bushing is about to engage the master bushing.

and the driller lowers the newly added joint and kelly until the kelly drive bushing engages the master bushing (fig. 151). The driller starts the pump, begins rotating, and lowers the bit back to bottom to continue making hole. Crew members make a connection each time the kelly is drilled down—each time the bit makes about 30 feet of hole. Near the surface, where the drilling is usually easy, they may make several connections while they are on tour.

Making a connection on a rig with a top drive is similar to making a connection on a rig that uses a rotary table. The driller simply raises the top drive a small amount to pick up the drill string. Crew members set the slips, which suspends the drill string in the hole. Note that although the rotary table is not used to rotate the drill string, it still provides a place for the crew to set the slips. Further, should the top drive fail, the driller can rotate the string and bit with the rotary table. With the drill string suspended by the slips, the driller actuates a control that tilts the *elevators* on the top drive. (Elevators are a special set of clamps that crew members latch around the tool joint of the drill pipe. Elevators grip the drill pipe joint and allow the driller to raise and lower the joint.) Tilting the elevators allows crew members to latch the elevators onto the joint in the mousehole (fig. 152B). The driller then picks up the joint from the mousehole and straightens the elevators.

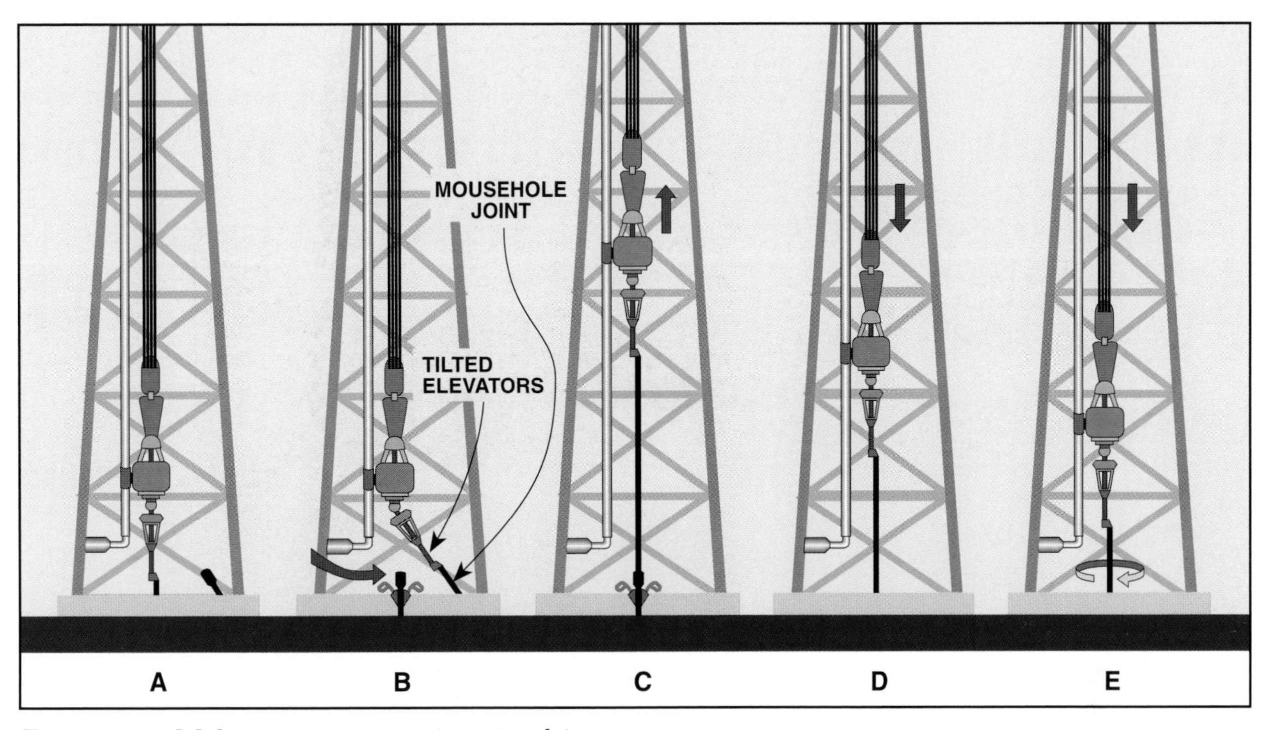

Figure 152. Making a connection using a top drive

With the elevators vertical, the crew stabs the new joint into the drill string suspended in the rotary table. The driller uses a built-in wrench in the top drive to spin up the joint and torque it to final tightness. Or, if desired, the crew can use tongs to buck up the joint. The driller then starts the mud pump, begins rotating the string and bit, and lowers the bit to bottom to continue drilling. Figure 152 (A–E) shows the steps in the operation.

At a predetermined depth, perhaps as shallow as a few hundred feet (metres) to as deep as two or three thousand feet (metres), drilling stops. Drilling stops because crew members drill this first part of the hole—the *surface hole*—only deep enough to get past soft, sticky formations, gravel beds, freshwater-bearing formations, and the like that lie relatively near the surface. At this point, crew members remove (*trip out*) the drill string and bit from the hole. They trip out the drill string and bit so that they can run casing into the hole. Once they cement the casing in place, the casing protects the formations that lie behind it and prevents fluids in the formations from migrating from one formation to another. For example, a saltwater zone could contaminate a freshwater zone if the operator did not case and cement the hole. The casing also protects these shallow zones from being contaminated by the drilling mud used to drill the hole below the casing.

TRIPPING OUT WITH A KELLY SYSTEM

To trip out (to remove the drill stem from the hole) on a rig with a kelly system, crew members set the slips around the drill stem, break out the kelly, and set it, the kelly drive bushing, and the swivel back in the rathole (fig. 153). Still attached to the bottom of the hook are the elevators. The driller lowers the traveling block and elevators down to the point where crew members can latch the elevators onto the pipe (fig. 154). The driller raises the traveling block, thus raising the elevators and pipe, and the floorhands remove the slips.

Figure 153. The kelly and swivel (with its bail) are put into the rathole.

Figure 154. Crew members latch elevators to the drill pipe tool joint suspended in the rotary table.

Meanwhile, the derrickman, using a safety harness and climbing device, has climbed up the mast or derrick to the *monkeyboard*. The monkeyboard is a small working platform on which the derrickman handles the top of the pipe. As the driller raises the pipe to the derrickman's level, the derrickman pulls the top of the pipe back into the *fingerboard* (fig. 155).

Figure 155. The derrickman has just placed the upper end of a stand of drill pipe between the fingers of the fingerboard.

The fingerboard, as the name implies, has several metal projections (fingers) that stick out to form slots into which the derrickman places the top of the pipe. When the floorhands move the pipe off to one side of the rig floor and set it down (fig. 156), the derrickman unlatches the elevators and prepares to receive the next stand of pipe. The floorhands do not break out (disconnect) every single joint of drill pipe or drill collar one at a time. Instead, they usually pull two or three joints at a time. So, although they put pipe into the hole one joint at a time when drilling, they pull it out two or three joints at a time.

Figure 156. The floorhands set the lower end of the stand of pipe off to one side of the rig floor.

Two or three joints together constitute what is termed a "stand." Crew members pull pipe out of the hole in stands to save time. If three joints comprise a stand, and that is the usual case, then the stand is sometimes called a "triple," or a "thribble" (although you don't hear this term much anymore). If two joints make up stand, it is called a "double." In a few cases, crew members may pull four-joint stands; in such a case, they pull "quadruples," or "fourbles." The height of the mast or derrick determines whether the crew pulls doubles, thribbles, or fourbles. Because the surface hole is usually relatively shallow, it does not take crew members very long to get all the drill stem and bit out of the hole.

TRIPPING OUT WITH A TOP-DRIVE UNIT

Tripping the drill string out of the hole with a top drive is much like tripping out with a kelly-and-rotary-table system. The main difference, of course, is that the top drive does not use a kelly and swivel. Thus, the rig has no need of a rathole. Instead, the driller simply unscrews the top drive's drive shaft from the drill string after the floorhands suspend the drill string in the hole with the slips. They use the top drive's built-in elevators to raise the pipe out of the hole and they usually use regular tongs to loosen the joints. The driller spins out the joints with the top drive. The floorhands then set the stands back in the mast or derrick while the derrickman handles the upper end of the stands on the monkeyboard.

Figure 157. A top view of an automatic pipe-handling device manipulating a stand of drill pipe

TRIPPING OUT WITH A PIPE RACKER

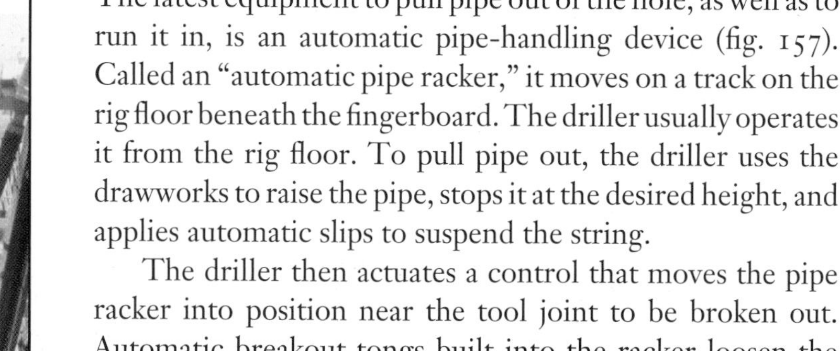

The latest equipment to pull pipe out of the hole, as well as to run it in, is an automatic pipe-handling device (fig. 157). Called an "automatic pipe racker," it moves on a track on the rig floor beneath the fingerboard. The driller usually operates it from the rig floor. To pull pipe out, the driller uses the drawworks to raise the pipe, stops it at the desired height, and applies automatic slips to suspend the string.

The driller then actuates a control that moves the pipe racker into position near the tool joint to be broken out. Automatic breakout tongs built into the racker loosen the joint and a built-in automatic spinner spins the joint apart. The operator retracts the racker's tongs and extends two arms from the racker. One of the arms grips the pipe near the rig floor and the other grips it higher up, just beneath the fingerboard. Gripping the pipe and moving the racker in its track sets the bottom of the stand or joint on the rig floor and places the top of the stand or joint into the fingerboard.

With the stand or joint set back in the mast or derrick, the driller retracts the racker's arms and moves the racker into position for the next stand or joint to be pulled from the hole. The driller extends the racker's automatic tongs and breaks out the next stand or joint of pipe. The racker then sets back the stand or joint in the fingerboard. The driller repeats this process until all the pipe is out of the hole.

RUNNING SURFACE CASING

Once the drill stem and bit are out of the hole, the casing crew moves in to do its work (figs. 158–163). Because the drilling crew just drilled the surface hole, the first string of casing run into the hole is called "surface casing," or "surface pipe."

Figure 158. A casing crew member cleans and inspects the casing as it lies on the rack next to the rig.

Figure 159. Casing threads must be clean, dry, and undamaged before they are run into the hole.

Figure 160. A joint of casing being lifted onto the rig floor

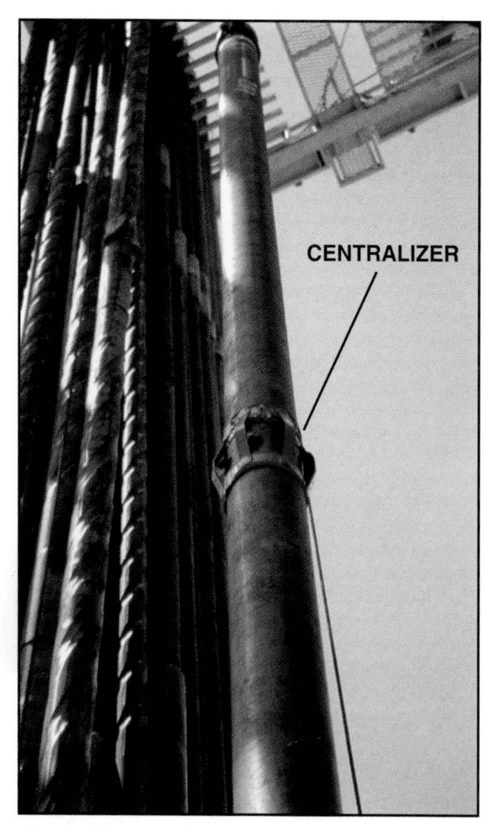

Figure 161. A joint of casing suspended in the mast; note the centralizer.

Figure 162. Casing elevators suspend the casing joint as the driller lowers the joint into the casing slips, or spider. The spider suspends the casing string in the hole.

Figure 163. Working from a platform called the "stabbing board," a casing crew member guides the casing elevators near the top of the casing joint.

Surface casing is usually large in diameter—perhaps 20 inches (508 millimetres) or more. Casing is strong steel pipe. Running casing into the hole is very similar to running drill pipe, except that the casing diameter is usually much larger and thus requires special elevators, slips, and tongs to fit it.

Also, the casing crew sometimes installs *centralizers* and *scratchers* on the outside of the casing before they lower it into the hole (fig. 164). The crew attaches centralizers around the outside of the casing joints and, because the centralizers have bowed springs, they keep the casing centered in the hole after the crew lowers it in. Ideally, casing should not come into contact with the walls of the hole. If it does, cement may not be able to flow into the area between the wall of the hole and the outside of the casing. Consequently, a *void* in the cement may occur, which could allow fluids to flow outside the casing (fig. 165). Fluid flow behind the casing is not desirable because contamination can occur. For example, salt water from one formation could flow into another formation containing fresh water and pollute it.

Scratchers also come into play when the casing is cemented. The idea is that if the driller moves the casing up and down or rotates it (depending on scratcher design), the scratchers remove the wall cake formed by the drilling mud and the cement will thus be able to bond better to the hole.

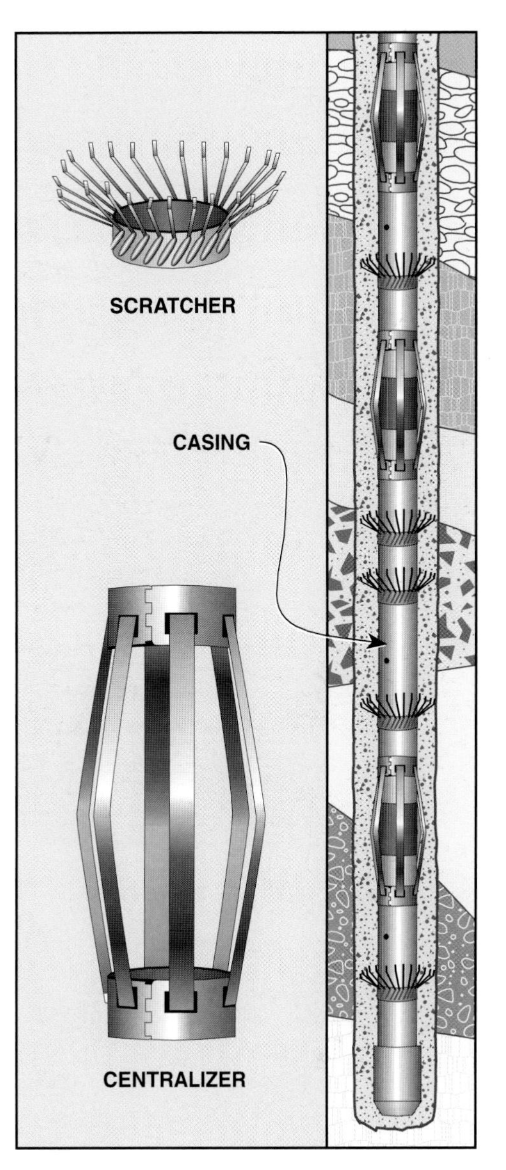

Figure 164. Crew members install scratchers and centralizers at various points in the casing string.

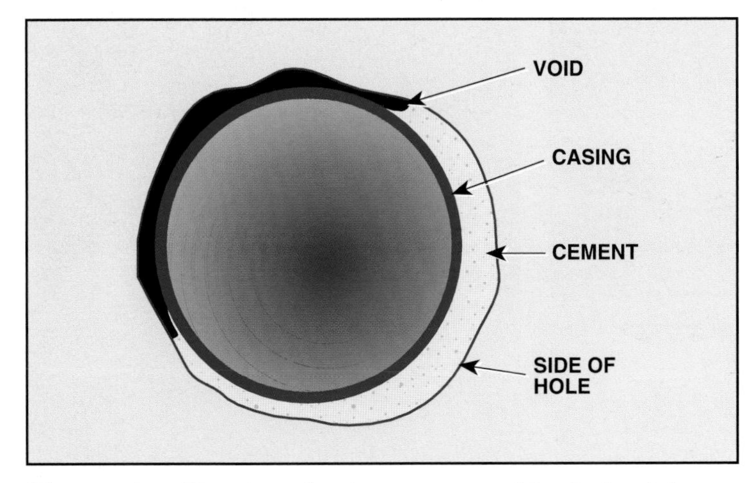

Figure 165. Top view of casing not centered in the borehole; a cement void exists where the casing contacts the side of the hole.

Other casing accessories include a *guide shoe*, which is a heavy steel-and-concrete fitting that the casing crew attaches to the bottom of the first joint of casing to go into the hole (fig. 166 A, B). The guide shoe helps guide the casing past small ledges or debris in the hole. Figure 167 is a schematic of the casing and accessories attached to it. Besides the guide shoe, another accessory is a *float collar* (fig. 168). The crew installs a float collar a couple of joints from bottom. The float collar keeps mud in the hole from entering the casing as the crew lowers the casing into the hole. Just as a ship floats in water, casing floats in a hole full of mud, if most of the mud is kept out of the casing. The float collar's valve keeps mud from entering the casing as the crew lowers it into the mud-filled hole. This buoyant effect helps relieve some of the weight carried by the mast or derrick.

Figure 166A. Crew members lift the heavy steel-and-concrete guide shoe.

Figure 166B. The guide shoe is made up on the bottom of the first joint of casing to go into the hole.

a crew
cemen
slurry
the mu
slurry
stops,
a mem
then g
last fev
guide s
the hol
space.
As
membe
167). A
membr
cement
ment fl
lated d
casing
cement
Co
the disp
slurry
Soon, t
float co
the pun
and in t
ment fl
Aft
moves
wait a s
referre
nounce
several,
ture, an
Aft
to ensu
membe
compa
problen
of the p
cement

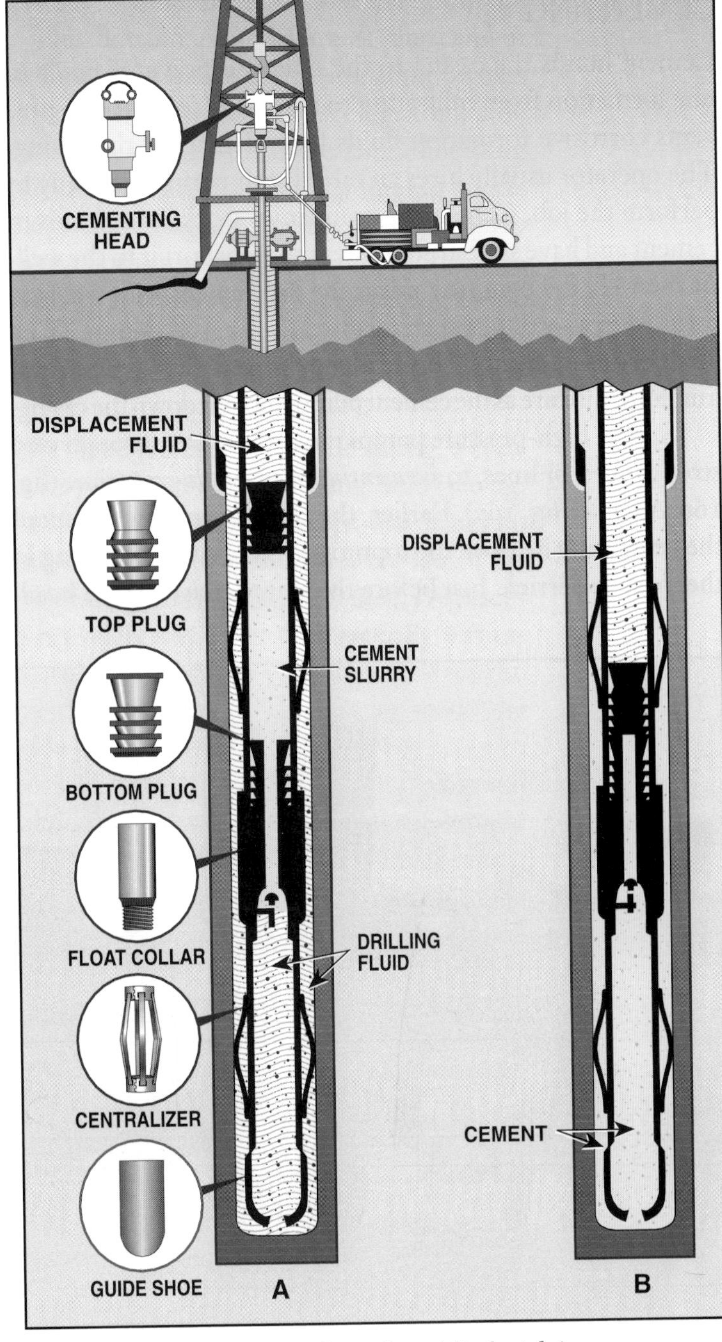

Figure 167. Cementing the casing; (A) the job in progress; (B) the finished job

Figure 168. Crew members install a float collar into the casing string.

TRIPPING IN

Once a good cement job is obtained, drilling can resume. The operator selects a bit that can go inside the surface casing. If, for example, the crew ran 13⅜-inch (346-millimetre) casing inside a 17½-inch (200-millimetre) hole, then they would probably run a 12¼-inch (311.2-millimetre) bit into the casing to drill the next stage of the hole. As before, crew members make up the new bit on the end of a drill collar and run it into the hole on more drill collars and enough drill pipe to get the bit to bottom. This process is known as "tripping in."

To trip in, the rotary helpers stab a stand of pipe into the suspended stand (fig. 170). They then spin up the joint with a spinning wrench (fig. 171). With the pipe spun up, crew members use the tongs to buck up the joint to final tightness (fig. 172).

Figure 170. To trip in, crew members stab a stand of drill pipe into another.

Figure 171. After stabbing the joint, crew members use a spinning wrench to thread the joints together.

Figure 172. After spin up, crew members use tongs to buck up the joint to final tightness.

Some rigs have an "Iron Roughneck™," which is a large machine that the floorhands use to spin up and buck up drill pipe joints (fig. 173). And some rigs have automatic pipe rackers, which run pipe into the hole with special gripping arms and a built-in spinning wrench and tongs.

Figure 173. An Iron Roughneck™ spins and bucks up joints with built-in equipment.

DRILLING AHEAD

Once the drill stem is back in the hole, the bit drills out the cement remaining in the lower part of the casing. It also drills out the guide shoe and begins drilling into the formation below the casing. From this point, the rig and crew may drill the hole to a formation that contains oil and gas. Or they may drill to another predetermined depth and temporarily stop. Whether they drill to final depth below the surface casing or to an intermediate depth depends on the nature of the formations the bit drills.

Some wells, especially deep ones, usually encounter formations that are easily controlled by using a suitable drilling fluid. Later, however, as the borehole drills into a deeper oil and gas formation, the drilling fluid used to control the upper zones is not suitable for the productive formation. The drilling fluid could damage the producing zone so badly that the operator could not withdraw the hydrocarbons from the zone. To avoid such a pitfall, the operator plans the well to be drilled to an intermediate depth above the *pay zone* (the productive formation). The drilling crew uses drilling fluid formulated to control the formations to the intermediate depth. Then, they stop drilling, *come out of the hole* (pull the bit and drill stem from the well), and run and cement casing. The casing and cement seal off the intermediate part of the hole so that the formations neither affect nor are affected by subsequent drilling operations.

When drilling the part of the hole below the surface casing, crew members probably won't make connections as frequently as they did when drilling the surface hole. Deeper formations tend to be hard and drilling is thus slower. Usually, at some point, the bit dulls and crew members have to replace it. To do so, they make a *round trip*. They trip out the dull bit and drill string and install a new bit. Then they trip in the new bit, the drill collars, and the drill pipe and resume drilling. Several round trips may be necessary before the hole reaches an intermediate or final depth.

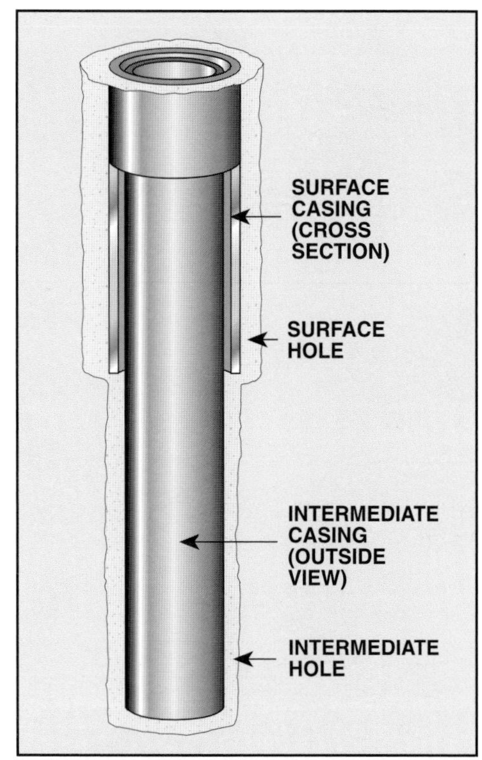

Figure 174. Intermediate casing is run and cemented in the intermediate hole.

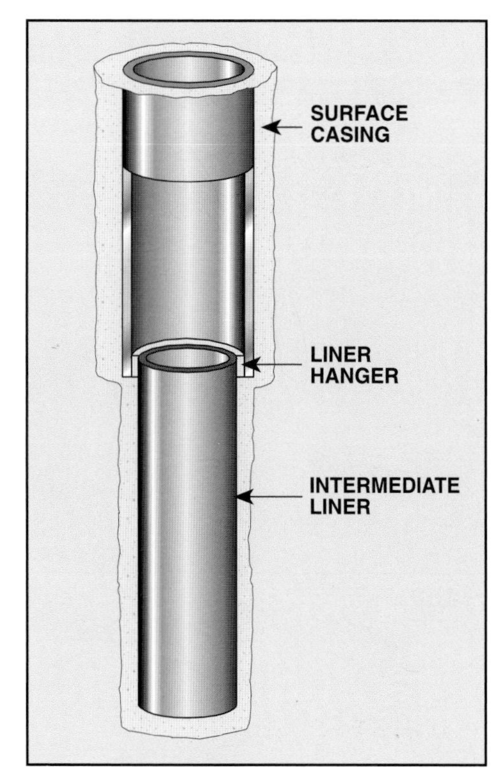

Figure 175. Intermediate liner is hung in the surface casing.

RUNNING AND CEMENTING INTERMEDIATE CASING

At this point, let's assume that the crew on our example well has drilled an intermediate hole. The operator then calls in a casing crew and cement company to run and cement casing through the surface casing and to the bottom of the intermediate hole. This casing string is the *intermediate string* (fig. 174).

The crew drilled the intermediate hole of our model well with a 12¼-inch (311.2-millimetre) bit, so the intermediate casing must fit inside this hole and leave room for cement. A couple of casing sizes are available, but let's say our operator picks one with an outside diameter of 9⅝ inches (244.5 millimetres). The crews run and cement this intermediate string of casing with the same equipment and techniques they used to run surface casing.

The operator may, however, opt to run an intermediate *liner* instead of casing. A liner is the same as casing except that it is shorter. A casing string runs from the bottom of the hole all the way to the surface. A liner string, on the other hand, runs from the bottom of the hole but stops a short distance inside the casing string above it. If crew members run an intermediate liner, they hang it in the bottom of the surface casing (fig. 175). The cementing crew then cements it back into the surface casing. Because they are shorter, liner strings save the operator money.

DRILLING TO FINAL DEPTH

The final part of the hole, whether crew members drill it through surface casing or through one or more intermediate casing or liner strings, is what the operator hopes will be the production hole. In the case of our example well, crew members make up a bit that fits inside the 9⅝-inch (244.5-millimetre) casing. Let's say the bit is 7⅞ inches (200 millimetres). Crew members trip in this bit, drill out the intermediate casing or liner shoe, and head for what everyone hopes is pay dirt—a formation that produces enough oil and gas to make it economically feasible for the company to complete the well. The question the operator faces is, "Does this formation contain enough oil or gas to make it worthwhile to run the final production string of casing or liner and complete the well?"

Formation Evaluation

11

Determining whether a formation contains oil and gas falls under the realm of formation evaluation. Formation evaluation includes the activities the operator does to test a formation for hydrocarbons. The operator must not only know whether hydrocarbons exist, but also whether they exist in ample amounts. A hole may penetrate a formation that contains hydrocarbons; however, if the formation does not contain enough hydrocarbons for the operating company to get its monetary investment back, the company may declare the hole to be dry. Methods of formation evaluation include examining cuttings and drilling mud, well logging, drill stem testing, and coring.

EXAMINING CUTTINGS AND DRILLING MUD

Several techniques are available to help the operator decide whether to complete the well. One of the simplest is looking at the cuttings the drilling mud carries from the bottom of the hole (fig. 176). A geologist can test the cuttings to determine whether they contain hydrocarbons. The mud logger, using various kinds of detection equipment, can also spot hydrocarbons in the drilling mud. An operator probably would not decide to complete or abandon a well using only information from cuttings and mud returns. Careful examination of them, however, can indicate whether the well is likely to produce.

Figure 176. A handful of cuttings made by the bit

WELL LOGGING

Well logging is a widely used evaluation technique. Many kinds of logging tool are available. Some measure and record natural and induced nuclear, or radioactive, attributes of a rock. Others measure and record the way in which formations respond to electric current. Another log measures and records the speed with which sound travels through a formation. These are only a few of many logs available to the operator. By interpreting the recordings, or logs, the operator can usually tell if the well will be a producer.

The operator calls the logging company to the well while the drilling crew trips out the drill string. From a portable laboratory, truck-mounted for land rigs or in a small cabin on offshore rigs, the well loggers lower logging tools into the well on wireline (fig. 177). They lower the tools to bottom and then slowly reel them back up. When activated, the tools measure formation properties.

Figure 177. Logging personnel run and control logging tools by means of wireline from this laboratory on an offshore rig.

The tools transmit the data they gather to the truck or logging shack. There, special recorders and computers store the information. For on-site evaluation, computers in the portable laboratory print the data (fig. 178). These logs give the operator a first look at what a formation may yield. For thorough evaluation, the portable lab can transmit the log's data to powerful computers located at central processing facilities. By carefully examining well logs, the operator can determine whether to complete the well. Well logs not only indicate the presence of oil and gas, they also indicate how much may be there.

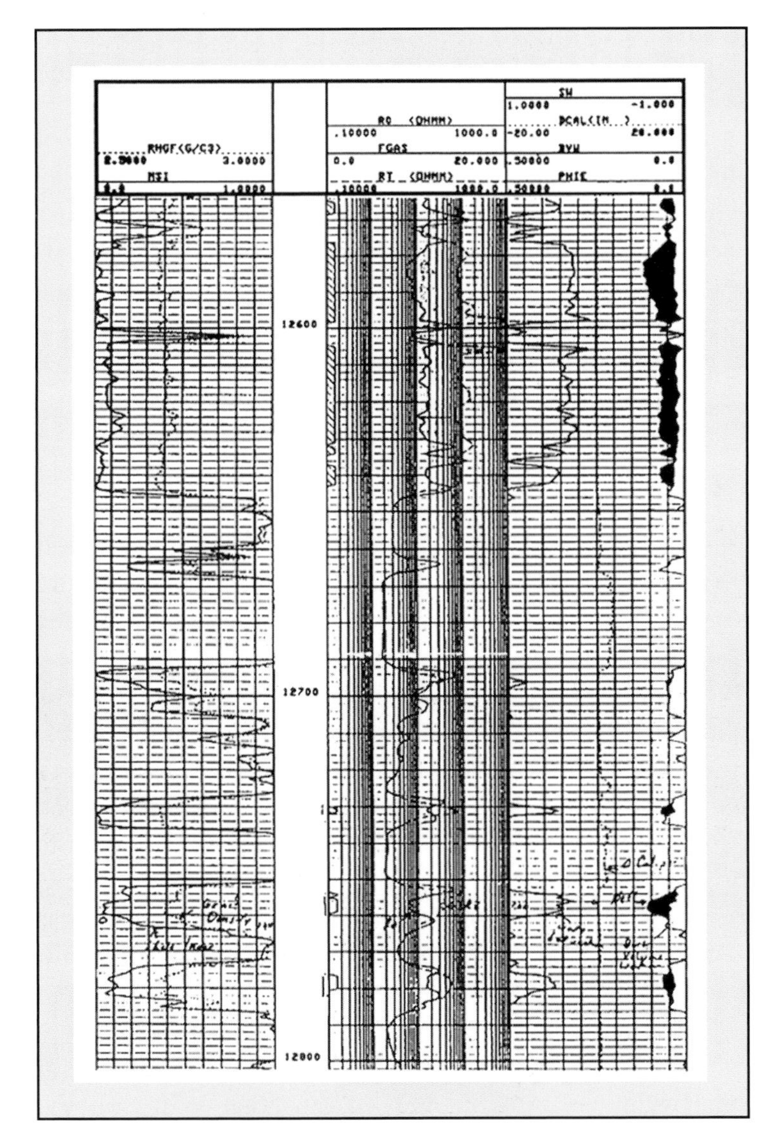

Figure 178. A well-site log is interpreted to give information about the formations drilled.

During drilling, the operator can run *logging while drilling (LWD)* tools in the drill stem. These instruments incorporate sophisticated electronic devices that sense, transmit, and record formation characteristics as the bit drills ahead. The LWD tool transmits formation information on a pulse the tool creates in the drilling mud. Much as radio waves transmit sound information through air, mud pulses transmit formation information to computers on the surface. The computers analyze and display the information in readouts that experts on the site can interpret and evaluate.

DRILL STEM TESTING

To further determine the potential of a producing formation, the operator may order a drill stem test, or DST (say "dee-ess-tee"). The DST crew makes up the test tool on the bottom of the drill stem then lowers the tool to the bottom of the hole. The crew applies weight to the tool to expand a hard-rubber sealing element called a packer. The expanded packer seals the hole above the packer. The DST crew then opens ports in the tool that lie below the packer. Opening the ports exposes recorders to pressure. Analysis of the downhole pressures indicates whether the well can produce.

CORING

In some cases, the operator may wish to examine directly a formation sample that is larger than the cuttings. If so, a core sample is ordered. Two coring methods are available. In one, the drilling crew makes up a core barrel and runs it to bottom. When the driller rotates the core barrel, it cuts a cylinder, or core, of rock. The core is often several inches (or millimetres) in diameter and several feet (or metres) long. As the core barrel cuts the core, the core moves into a tube in the barrel. After the desired length of core is cut, the crew trips out the drill string and the core barrel. At the surface, the core is removed and shipped to a laboratory for thorough analysis.

The second coring method uses a sidewall sampler. In this method, the crew lowers the sampling device to the desired depth. The driller actuates a switch to make the sampler fire an explosive charge. The explosion rams several small cylinders into the wall of the hole. Wires on the cylinders keep them attached to the sampler. Then when the crew retrieves the sampler, the cylinders, along with bits of the formation, follow the sampler to the surface. Sidewall samplers can obtain up to thirty small samples from any depth.

12 Completing the Well

The operating company carefully considers the data it obtains from the tests. Then it decides whether to set production casing or liner and complete the well or to plug and abandon it. If the company decides to abandon it, the hole is *dry*. Dry in the sense of an oil or gas well means the well cannot produce oil or gas in commercial quantities. Some oil or gas may be present but not enough to justify the expense of completing the well. If the well is dry, the operator hires a cementing company to place several cement plugs in the well to seal it permanently.

SETTING PRODUCTION CASING

If the operator decides to set production casing, a supplier brings it to the well. For the final time, the casing and cementing crews run and cement a string of casing in the well. In the case of our model well, the crew could run 5-inch (127-millimetre) casing in the 7⅞-inch (200-millimetre) hole. Keep in mind that the operator may elect to set a liner string. As you recall, a liner string is the same as a casing string except that a liner does not run all the way to the surface. Instead, the casing crew hangs it inside a previously run casing or liner.

Usually, the casing and cementing crews set and cement the production casing or liner through the pay zone. The drilling crew drills the hole so that it goes all the way through the producing horizon and stops a short distance below. Then the casing crew runs the production string almost to the bottom of the hole. (It leaves a little room beneath the guide shoe to allow cement to flow out of the casing.) The production casing or liner and the cement actually seal off the producing zone. At this point, the drilling rig and crews are finished with their job: they have drilled, cased, and cemented the well to the depth specified by the drilling contract. Their only remaining job is to disassemble the rig (rig down) and move it to the next drilling location.

PERFORATING

The operator is not through, however. Because the production string and the cement seal the producing zone, the operator has to provide a way for oil and gas to get from the formation and into the well. Usually, the operator hires the services of a completion rig, which is a relatively small portable rig whose crews perform the final operations required to bring the well into production (fig. 179).

Figure 179. This small rig is a well servicing and workover unit. The operator often employs such units to complete a well.

Figure 180. Perforations (holes)

One important task is to perforate the well. A special gun shoots several relatively small holes in the casing. It makes them in the side of the casing opposite the producing zone. These holes, or *perforations* (fig. 180), pierce the casing or liner and the cement around the casing or liner. The perforations go through the casing and the cement and a short distance into the producing formation. Formation fluids, which include oil and gas, flow through these perforations and into the well.

The most common perforating gun uses shaped charges, similar to those used in armor-piercing shells. Several high-speed, high-pressure jets of gas penetrate the steel casing, the cement, and the formation next to the cement. A perforating specialist installs the charges in the special gun and lowers it—usually on wireline, rather than drill pipe—into the well to the desired depth. The depth can be determined by running a collar locator log, which identifies the depth of each casing collar. By comparing the log with the overall number and length of the casing joints, the operator can accurately determine the depth. Once at the desired depth, the perforating specialist fires the gun to set off the charges (fig. 181). After the gun makes the perforations, the perforating specialist retrieves it.

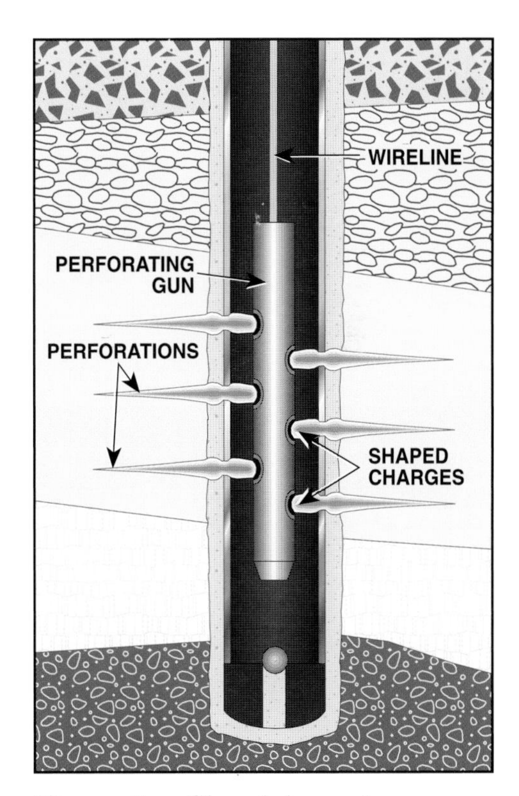

Figure 181. Shaped charges in a perforating gun make perforations.

RUNNING TUBING AND INSTALLING THE CHRISTMAS TREE

After the well is perforated, oil and gas can flow into the casing or liner. Usually, however, the operator does not produce the well by allowing hydrocarbons to flow up the casing or liner. Instead, the completion rig crew places small-diameter pipe called "tubing" inside the cased well. In fact, the operator sometimes runs tubing into the well before perforating it. In such cases, the perforating gun is lowered through the tubing to the required depth.

Tubing that meets API specifications has an outside diameter that ranges from 1.050 inches (26.7 millimetres) to 4½ inches (114.3 millimetres). Seven sizes between the two extremes are also available. As it does with casing, the crew commonly uses couplings to join tubing, although an integral-joint tubing is available that allows the crew to make up joints without using couplings.

Manufacturers also supply *coiled tubing*. Coiled tubing is a continuous length—it does not have joints—of flexible steel pipe that comes rolled on a large reel. Operators have completed wells over 20,000 feet (6,000 metres) deep with coiled tubing. Special equipment placed at the top of the well allows crew members to insert, or inject, the tubing into the well as they unwind it from the reel (fig. 182). The main advantage of coiled tubing is that crew members do not have to connect several single joints of tubing when installing the string. Consequently, coiled tubing takes considerably less time to run.

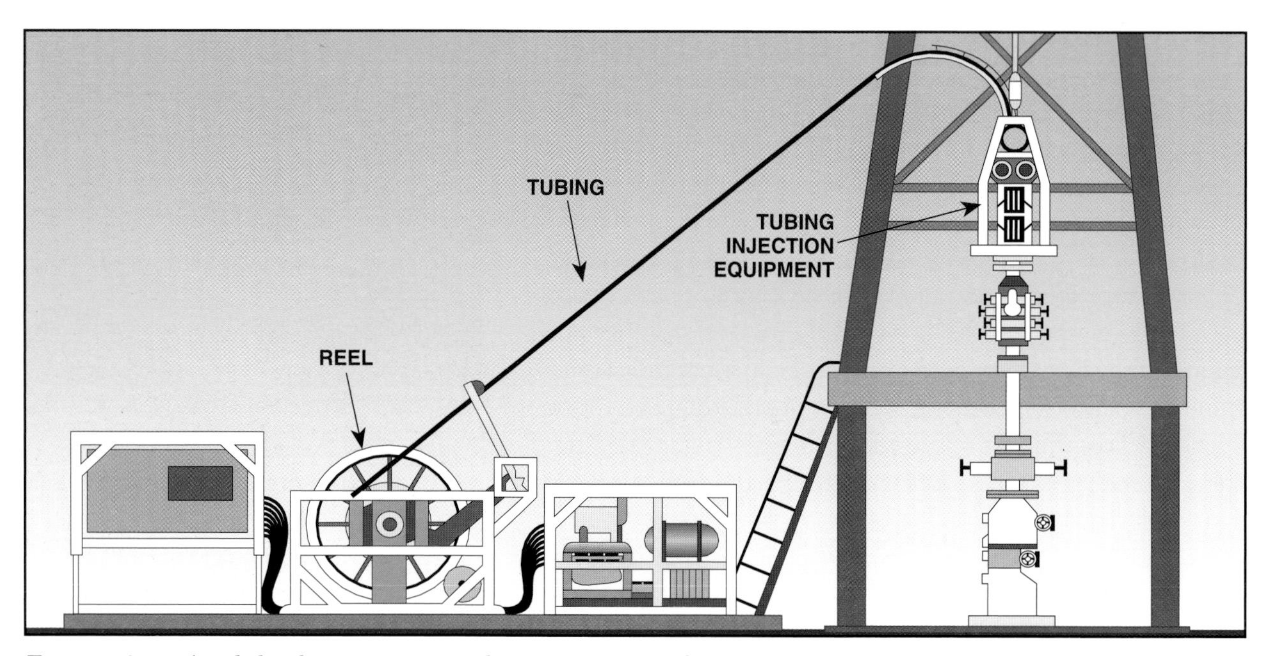

Figure 182. A coiled-tubing unit runs tubing into the well from a large reel.

Whether using jointed or coiled tubing, the operator usually produces a well through a tubing string rather than through the casing for several reasons. For one thing, the crew does not cement a tubing string in the well. Accordingly, when a joint of tubing fails, as it almost inevitably will over the life of a well, the operator can easily replace the failed joint or joints or, in the case of coiled tubing, remove and repair or replace the failed area. Since casing is cemented, it is very difficult to replace.

For another thing, tubing allows the operator to control the well's production by placing special tools and devices in or on the tubing string. These devices allow the operator to produce the well efficiently. In some cases, the operator can produce the well only by using a tubing string. Casing does not provide a place to install any tools or devices that may be required for production. In addition, the operator installs safety valves in the tubing string. These valves automatically stop the flow of fluids from the well if damage occurs at the surface.

Finally, tubing protects the casing from the corrosive and erosive effects of produced fluids. Over the life of a well, reservoir fluids tend to corrode metals with which they are in contact. By producing fluids through the tubing, which the operator can easily replace, the casing, which is not so easy to repair or replace, is preserved.

Crew members usually run tubing into the well with a sealing device called a "packer." They install the packer on the tubing string and place it at a depth slightly above the casing perforations. The end of the tubing is left open or is perforated and extends to a point opposite the perforations in the casing. The packer expands and grips the wall of the production casing or liner. When expanded, the packer seals the annular space between the tubing and the casing above the perforations. The produced fluids flow through the perforations and into the tubing string. The packer prevents them from entering the annular space, where they could eventually corrode the casing.

After the crew runs the tubing string, the operator has a crew install a collection of fittings and valves called a Christmas tree (fig. 183) on top of the well. Tubing hangs from the tree so the well's production flows from the tubing and into the tree. Valves on the Christmas tree allow the operator to control the amount of production or to shut in the well completely to stop it from producing. They also allow the operator to direct the flow of production through various surface lines as required. In addition, a special safety valve on the tree automatically shuts in the well if the tree is damaged. This automatic shut-in valve prevents reservoir fluids from flowing onto the surface if damage occurs. Usually, once the crew installs the Christmas tree, the well is complete.

Figure 183. This collection of valves and fittings is a Christmas tree.

ACIDIZING

Hydrocarbons sometimes exist in a formation but cannot flow readily into the well because the formation has very low permeability. If the formation reacts favorably to acid, acidizing may improve flow. An acidizing service company can pump anywhere from 50 to thousands of gallons (or litres) of acid down the well's tubing. The acid, to which the acidizing company adds a chemical to prevent it from corroding the tubing, enters the perforations and contacts the formation. Continued pumping forces the acid into the formation, where it etches channels. These channels provide a way for the formation hydrocarbons to enter the well through the perforations.

FRACTURING

Another treatment that may improve flow is *fracturing*. A fracturing service company pumps a specially blended liquid down the well's tubing and into the perforations. The pumps develop a great deal of pressure at very high rates of flow. Continued pumping causes the formation to split, or fracture, much as a steel wedge causes a log to split. The fracturing crew adds a finely graded sand or similar material (a *proppant*) to the fracturing fluid. The proppant enters the fracture in the formation. When the fracturing crew stops the pumps, the pressure dissipates. With the pressure gone, the fracture tries to close. The proppants, however, hold the fracture open. This propped-open fracture provides a passage for hydrocarbons to flow into the well.

Special Operations

13

For our purposes, special drilling operations include directional drilling, fishing, and well control. *Directional drilling* is intentionally drilling the hole off-vertical for various reasons. *Fishing* is the operation crew members implement to retrieve an object in the wellbore that doesn't belong there and impedes drilling. *Well control* is the techniques crew members use to regain control of the well should formation fluids inadvertently enter the well.

DIRECTIONAL AND HORIZONTAL DRILLING

Often, the drilling crew tries to drill the hole as straight as possible. Sometimes, however, the operator wants the hole to be drilled at a slant. One area where operators use slant, or directional, drilling is offshore. From a permanent platform that the operator installs over the drilling site, the crew must drill several wells to exploit the reservoir properly. To do so, crew members drill several *directional wells* (fig. 184). The crew may drill only the first well vertically; it drills every other well directionally.

To drill a typical directional well, the crew members drill the first part of the hole vertically. Then they *kick off*, or *deflect*, the hole so that the bottom may end up hundreds of feet or metres away from its starting point on the surface. By using directional drilling, the crew can drill forty or more wells into the reservoir from a single platform.

Another use of directional drilling is *horizontal drilling*. An operator can better produce certain reservoirs with horizontal drilling. The drilling crew drills the well vertically to a point above the reservoir. Then it deflects the well and increases the angle until it reaches 90 degrees, or horizontal (see fig. 112). This horizontal hole penetrates the reservoir. When properly applied, one horizontal borehole can produce a reservoir better than several vertically drilled ones.

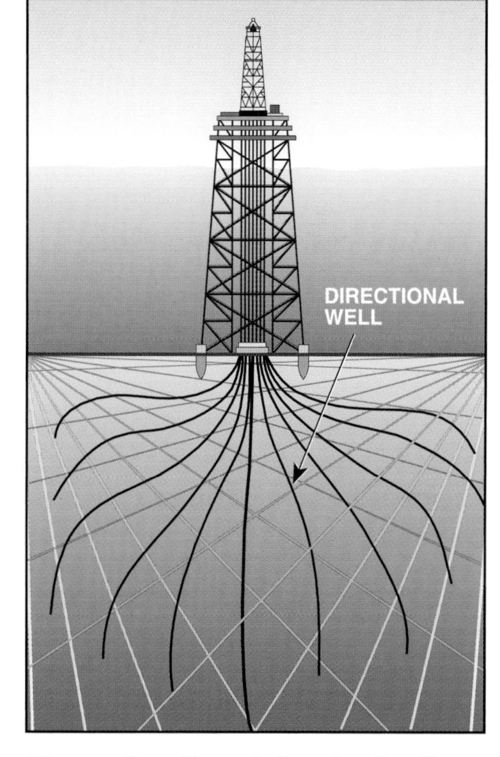

Figure 184. Several directional wells tap an offshore reservoir.

In horizontal and directional drilling, the crew can bend the drill stem to a high degree without breaking it because, first, the crew gradually deflects the hole from vertical. Usually, crew members deflect the hole over hundreds of feet (or metres) so that the bend is not sudden. Three to 10 degrees of deflection over 100 feet (or metres) is not an unusual amount. Second, the drill string is flexible. It is, after all, a hollow metal tube. The crew can bend it quite a lot without its breaking or permanently bending. In cases in which the hole needs to curve within a short distance, they use a special segmented pipe. Segmented pipe is very flexible and can bend a great deal without breaking.

The crew uses many tools and techniques to drill directionally. One tool is a downhole motor (fig. 185A), which crew members run with a bent sub (fig. 185B). A downhole

Figure 185A. A downhole motor laid on the rack prior to being run into the hole.

Figure 185B. A bent sub deflects the bit a few degrees off-vertical to start the directional hole.

motor is shaped like a piece of pipe. One type has turbine blades inside it. A turbine is like a series of fan blades stacked on top of each other on a shaft. Another type has a spiral-shaped steel shaft inside an elliptical opening. The bend in the sub is from 1 to 3 degrees. This small degree of bend allows the crew to run the tool into the hole without the tool's hanging up on the side of the hole. Although the bend is small, it starts the hole at an angle that the crew can increase as drilling progresses. With either tool, the crew makes up a bit on the bottom of the tool. Then it runs the bit and the tool to bottom, as usual.

When the tool reaches bottom, the crew turns the string to get the tool to face in the desired direction. In other words, crew members point the bend of the tool in the direction necessary to make the hole go where they want it to. They usually do so by using *measurement while drilling* (MWD) tools and techniques. MWD is similar to LWD. To use MWD, crew members place an MWD tool in the drill string as close to the bit as possible. As mud circulates through the string and past the MWD tool, the tool generates pulses in the mud. These pulses move up the drill string against the down flowing mud. Similar to the way in which radio waves carry music, voice, and other information, the pulses transmit directional and other data to the surface. Computers at the surface interpret the data and read it out to the directional operator.

To drill, the crew does not rotate the drill string. Instead, drilling mud flowing through the directional motor causes the turbine blades to turn, or the spiral shaft to turn, which rotates the bit. Once drilling gets under way, the MWD tool constantly sends the direction the hole is heading to the surface. It also transmits the angle of the hole.

FISHING

A *fish* is a piece of equipment, a tool, or a part of the drill string that the crew loses in the hole. Drilling personnel call small pieces, such as a bit cone or a wrench, "junk." Whenever junk or a fish exists in a hole, the crew has to remove it, or fish it out. Otherwise, it cannot continue to drill. Over the years, fishing crews have developed many ingenious tools and techniques to retrieve fish. For example, the crew can run an *overshot* into the hole to the fish. Crew members make up the overshot on drill pipe and lower the overshot over the fish. Grapples in the overshot latch onto the fish firmly. Then the crew pulls the overshot and attached fish out of the hole (fig. 186).

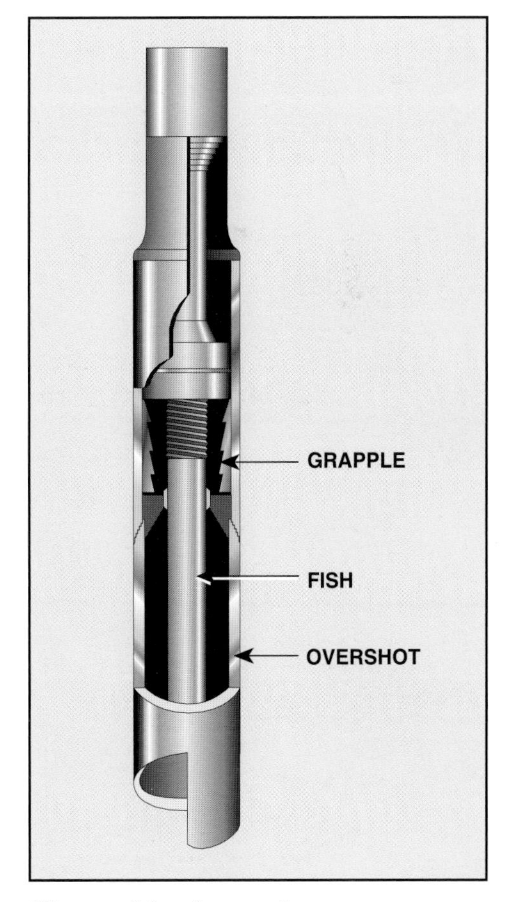

Figure 186. An overshot

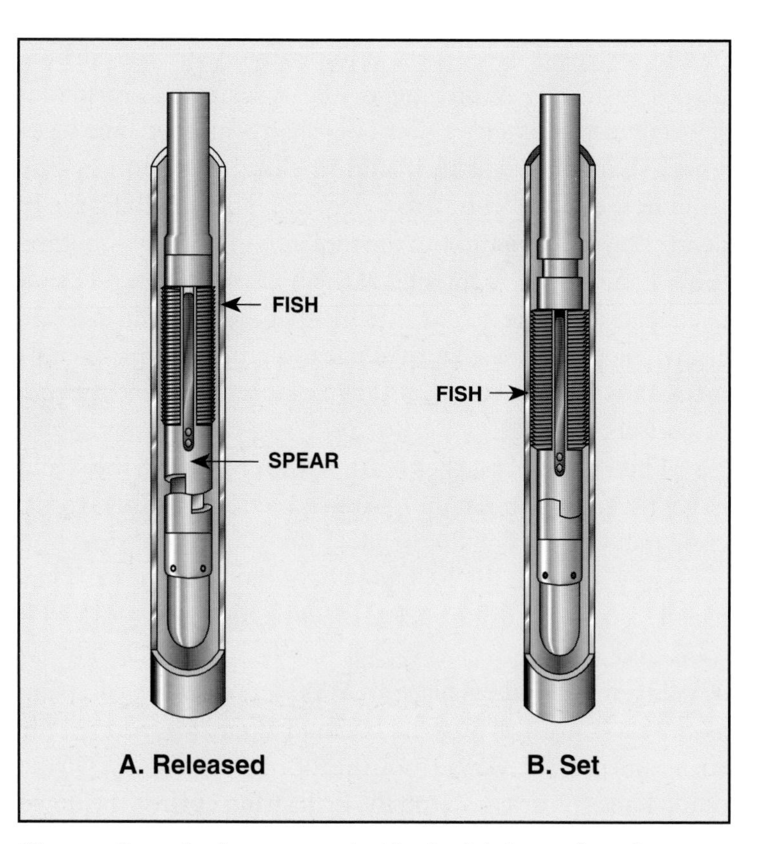

Figure 187. A. A spear goes inside the fish in a released position; B. once positioned, the spear is set and the fish removed.

Another fishing tool is a *spear* (fig. 187). Unlike an overshot, which the crew places over the fish, a spear grips inside the fish and allows the crew to retrieve it. Other fishing tools include powerful magnets and baskets. The crew uses them to fish for junk. Since no two fishing jobs are alike, manufacturers and fishing experts have developed many other fishing tools to meet the unique needs of fishing crews.

WELL CONTROL

As mentioned earlier, one vital job drilling fluid should do is keep formation fluids from entering the wellbore. If enough formation fluids enter the wellbore, drilling personnel say that the well "kicks." A kick, if not recognized and properly handled, can lead to a *blowout*. A blowout can be a catastrophic event (fig. 188). In many cases, fluids in the blowout ignite and reduce the rig to a melted pile of junk. Blowouts not only waste oil and gas, but also threaten human lives. Obviously, drilling crews take a great deal of care not to allow blowouts, and, in fact, not many occur. But, because a blowout is often a spectacular show and human lives are sometimes lost, a blowout often becomes a media event. Unfortunately, the impression may linger that blowouts are not the rarity they actually are. In reality, thousands of wells are drilled every year and very few of them blow out.

Figure 188. Fluids erupting from underground caught fire and melted this rig.

A hole full of mud that weighs the right amount—has the correct density—cannot blow out. But sometimes the unexpected occurs. Because a rig crew is only human, they can make an error and allow formation fluids such as gas, oil, or salt water to enter the hole. When formation fluids enter the hole—when a kick occurs—it makes it presence known by certain things that happen in the circulating system. For example, the level of mud in the tanks may rise above normal level, or mud may flow out of the hole even with the mud pump stopped. Alert drilling crews spot these anomalies (although the anomalies are sometimes subtle) and take steps to control the well and prevent a blowout.

When crew members discover a kick, they bring *blowout preventers*, or *BOPs* ("bee-oh-pees") into play. On land rigs and on offshore rigs that are not floaters, such as jackups, crew members *nipple up* (attach) the BOPs to the top of the well below the rig floor (fig. 189). The preventers are large, high-pressure valves capable of being remotely controlled.

Figure 189. A stack of blowout preventers (BOPs) installed on top of the well

When closed, they form a pressure-tight seal at the top of the well and prevent the escape of fluids. On floating offshore rigs, such as semisubmersibles and drill ships, crew members lower the BOPs to the top of the well on the seafloor (fig. 190).

Two basic types of blowout preventers are annular and ram. Crew members usually mount the *annular preventer* at the very top of the stack of BOPs. They call it an annular preventer because it seals off the annulus between the drill stem and the side of the hole. An annular BOP can also seal *open hole*—hole that has no pipe in it. Crew members typically mount two, three, or four ram-type BOPs below the annular preventer. *Ram preventers* get their name from the fact that the devices that seal off the well are large, rubber-faced blocks of steel that, when actuated, come together much like a couple of fighting rams butting heads. The two main kinds of ram preventers are *blind rams*, which seal off open hole, and *pipe rams*, which seal off the hole when drill pipe is in use. Normally, the driller closes the annular preventer first when the crew detects a kick. If it should fail, or if special techniques are required, the driller uses the ram-type preventers as a backup.

Figure 190. A subsea stack of BOPs being lowered to the seafloor from a floating rig

Closing in the well with one or more of the blowout preventers is only the first step in controlling the well. To resume drilling, crew members have to circulate the kick out and circulate in mud of the proper weight. To facilitate circulating a kick, crew members install a series of valves called the "choke manifold" (fig. 191). A choke is simply a valve with an opening the size of which a person can adjust

Figure 191. Several valves and fittings make up a typical choke manifold.

Figure 192. A remote-controlled choke installed in the choke manifold

(fig. 192). A choke operator, by using a remote control panel, can vary the size of the choke opening between fully open and fully closed (fig. 193). To circulate a kick out of the well and pump heavy mud in, the person operating the choke fully opens it, the driller starts the mud pump, and, as the kick fluids start moving up the hole, the choke operator reduces the choke opening to hold just enough back-pressure on the well to allow the mud and kick out but prevent further entry of formation fluid. Once the kick is out and the heavier mud in, the drilling crew makes a few checks to ensure that the well is back under control, and drilling operations resume.

Figure 193. This control panel allows an operator to adjust the size of the choke opening.

Rig Safety and Environmental Concerns

14

Drilling pioneers such as Drake, Uncle Billy, the Hamils, Lucas, and many others would undoubtedly be impressed by the progress made in drilling tools and techniques. What's more, they would also be impressed by the significant advances drilling contractors and operators have made in safeguarding personnel. Although rig safety may not be as glamorous as technical improvements, it is vitally important. That operators and contractors have taken great strides in personnel safety is borne out by the fact the accident rate on rigs is decreasing. Indeed, accidents have trended downward over the last several years. A look at IADC accident statistics for a recent year show that rig crews all over the world worked almost 200 million hours. Yet, there were just 1,001 *lost time accidents*. (A lost time accident is one that is serious enough to prevent the injured person from working the next scheduled workday.) It may be easier to fathom just how low this rate is if you consider that for every 200,000 hours rig personnel worked only one suffered an injury serious enough to prevent him or her from working the next day.

Part of the downward trend relates to training. Contractors and operators now consider training an essential part of preparing new workers for the rig. What's more, training is ongoing: not only are new personnel trained, but also experienced personnel at all levels receive advanced and refresher training on a regular basis. In addition to intensive training of rig personnel, contractors and operators have taken great steps in designing drilling rigs to be as safe a place to work as possible. For example, no contractor today would ever consider erecting a rig without adequate protective shrouds, or guards, on rig machinery. Steel covers over and around moving parts protect crew members from inadvertently contacting them. Personal protective gear that prevents or minimizes injury to the eyes, head, ears, and feet is standard apparel for everyone on the rig site. In addition, when handling particularly hazardous materials, such as caustic soda, additional protective gear is required. Climbing aids and fall protection equipment are also standard on today's rigs. Hand rails, guard rails, and nonskid surfaces on all walkways and passageways keep falls and slips to a minimum. What's more, signs, placards, and safety information alert personnel to potential rig hazards and provide information on avoiding illness or injury.

Protecting the environment from harm is another area in which contractors and operators have made great advances. Earlier, you read that contractors sometimes place nets over reserve pits to keep migratory water fowl from landing in them. Such action is only one of many steps contractors and operators take to protect the environment. Additional examples include installing plastic lining on reserve pits to prevent water or other materials from leaching into the soil, cleaning of oil-laden cuttings before they are disposed of, and, in especially sensitive areas, prohibiting any discharge onto the ground or into the water.

In many ways, today's rotary rigs are not that different from the rotary rigs of yore, such as the one the Hamils used to drill Spindletop. At the same time, however, modern rigs are considerably advanced. The industry has come a long way since the days of "wooden derricks and iron men." Granted, the basic name of the game is still putting a bit on bottom and turning it while circulating drilling fluid, but today's tools and techniques have evolved to make rotary rigs more efficient than ever. Steel has replaced that which used to be wood and modern steel alloys have replaced steel that used to break or wear out prematurely. Moreover, rig personnel are trained to work safer than ever before. Eventually, other forms of energy will supplant oil and gas, but, at least for now, the sight of a rotary drilling rig with its bit on bottom and turning to the right is not likely to disappear.

Glossary

A

abandon *v*: to cease producing oil and gas from a well when it becomes unprofitable or to cease further work on a newly drilled well when it proves not to contain profitable quantities of oil or gas.

accumulator *n*: see *blowout preventer control unit.*

acid fracture *v*: to part or open fractures in productive hard limestone formations by using a combination of oil and acid or water and acid under high pressure. See *formation fracturing.*

acidize *v*: to treat oil-bearing limestone or other formations with acid for the purpose of increasing production.

adjustable choke *n*: a choke in which the position of a conical needle, sleeve, or plate may be changed with respect to their seat to vary the rate of flow; may be manual or automatic. See *choke.*

afternoon tour *n*: on rigs that employ three 8-hour shifts, the work period that covers the afternoon and evening hours, such as from 3:00 pm to 11:00 pm. Also called evening tour.

air-actuated *adj*: equipment activated by compressed air, as are the clutch and the brake system in drilling equipment.

air drilling *n*: a method of rotary drilling that uses compressed air as the circulation medium.

air hoist *n*: a hoist operated by compressed air; a pneumatic hoist. Air hoists are often mounted on the rig floor and are used to lift joints of pipe and other heavy objects.

air tugger *n*: see *air hoist.*

American Petroleum Institute (API) *n*: oil trade organization (founded in 1920) that is the leading standard-setting organization for all types of oilfield equipment. It maintains departments of production, transportation, refining, and marketing in Washington, DC. It offers publications regarding standards, recommended practices, and bulletins. Address: 1220 L St., NW; Washington, DC 20005; (202) 682-8000.

angle of deflection *n*: in directional drilling, the angle at which a well diverts from vertical; usually expressed in degrees, with vertical being 0°.

annular blowout preventer *n*: a large valve, usually installed above the ram preventers, that forms a seal in the annular space between the pipe and the wellbore or, if no pipe is present, in the wellbore itself. Compare *ram blowout preventer.*

annular preventer *n*: see *annular blowout preventer.*

annular space *n*: the space between two concentric circles. In the petroleum industry, it is usually the space surrounding a pipe in the wellbore, or the space between tubing and casing, or the space between tubing and the wellbore; sometimes termed the annulus.

annulus *n*: see *annular space.*

anticline *n*: rock layers folded in the shape of an arch. Anticlines sometimes trap oil and gas.

area drilling superintendent *n*: an employee of a drilling contractor whose job is to coordinate and oversee the contractor's drilling projects in a particular region or area.

assistant driller *n*: a member of a drilling rig's crew whose job is to aid and assist the driller during rig operations. This person not only controls the drilling operation at certain times, but also keeps records, handles technical details, and, in general, keeps track of all phases of the operation. See *driller*.

assistant rig superintendent *n*: an employee of a drilling contractor whose job includes aiding the rig superintendent; in some cases, the assistant rig superintendent takes over for the rig superintendent during nighttime hours. Consequently, the assistant rig superintendent is sometimes called the night toolpusher. See *rig superintendent*.

automatic pipe racker *n*: a device used on a drilling rig to automatically remove and insert drill stem components from and into the hole. It replaces the need for a person to be in the derrick or mast when tripping pipe into or out of the hole.

B

back-in unit *n*: a portable servicing or workover rig that is self-propelled, using the hoisting engines for motive power. Because the driver's cab is mounted on the end opposite the mast support, the unit must be backed up to the wellhead. See *carrier rig, drive-in unit*.

back off *v*: to unscrew one threaded piece (such as a section of pipe) from another.

back up *v*: to hold one section of an object such as pipe while another section is being screwed into or out of it.

bail *n*: a curved steel rod on top of the swivel that resembles the handle, or bail, of an ordinary bucket, but is much larger. Sometimes, the two steel rods (the links) that attach the elevator to the hook are also called bails. *v*: to recover bottomhole fluids, samples, mud, sand, or drill cuttings by lowering a cylindrical vessel called a bailer to the bottom of a well, filling it, and retrieving it.

bailer *n*: a long, cylindrical container fitted with a valve at its lower end, used to remove water, sand, mud, drilling cuttings, or oil from a well in cable-tool drilling.

ball up *v*: to collect a mass of sticky consolidated material, usually drill cuttings, on drill pipe, drill collars, bits, and so forth.

barge *n*: a flat-decked, shallow-draft vessel, usually towed by a boat. A complete drilling rig may be assembled on a barge and the vessel used for drilling wells in lakes and in inland waters and marshes.

barge control operator *n*: an employee on a semisubmersible rig whose main duty is to monitor and control the stability of the rig. From a special work station on board the rig, this person controls the placement of ballast water inside the rig's pontoons to maintain the rig on even keel during all operations.

barge engineer *n*: see *barge control operator*.

barge master *n*: see *barge control operator*.

barite *n*: barium sulfate, $BaSO_4$; a mineral frequently used to increase the weight or density of drilling mud.

barium sulfate *n*: a chemical compound of barium, sulfur, and oxygen ($BaSO_4$). Also called barite.

barrel (bbl) *n*: a measure of volume for petroleum products in the United States. One barrel is the equivalent of 42 U.S. gallons or 0.15899 cubic metres (9,702 cubic inches). One cubic metre equals 6.2897 barrels.

barrels per day (bpd) *n*: in the United States, a measure of the rate of flow of a well; the total amount of oil and other fluids produced or processed per day.

bbl *abbr*: barrel.

bed *n*: a specific layer of earth or rock that presents a contrast to other layers of different material lying above, below, or adjacent to it.

bedrock *n*: solid rock just beneath the soil.

belt *n*: a flexible band or cord connecting and wrapping around each of two or more pulleys to transmit power or impart motion.

belt guard *n*: a protective grill or cover for a belt and pulleys.

bent sub *n*: a short cylindrical device installed in the drill stem between the bottommost drill collar and a downhole motor. Its purpose is to deflect the downhole motor off vertical to drill a directional hole.

bit *n*: the cutting or boring element used in drilling oil and gas wells. The bit consists of a cutting element and a circulating element. The cutting element is steel teeth, tungsten carbide buttons, industrial diamonds, or polycrystalline diamond compacts (PDCs).

bit breaker *n*: a heavy plate that fits in the rotary table and holds the drill bit while it is being made up in or broken out of the drill stem. See *bit*.

bit cutter *n*: the teeth of a bit.

bit pin *n*: the threaded element at the top of a bit that allows it to be made up in a drill collar or other component of the drill stem.

bit record *n*: a report that lists each bit used during a drilling operation, giving the type, the footage it drilled, the formation it penetrated, its condition, and so on.

bit sub *n*: a sub inserted between the drill collar and the bit. See *sub*.

blind ram *n*: an integral part of a blowout preventer, which serves as the closing element on an open hole. Its ends do not fit around the drill pipe but seal against each other and shut off the space below completely. See *ram*.

blind ram preventer *n*: a blowout preventer in which blind rams are the closing elements. See *blind ram.*

block *n*: any assembly of pulleys on a common framework; in mechanics, one or more pulleys, or sheaves, mounted to rotate on a common axis. The crown block is an assembly of sheaves mounted on beams at the top of the derrick or mast. The drilling line is reeved over the sheaves of the crown block alternately with the sheaves of the traveling block, which is raised and lowered in the derrick or mast by the drilling line.

blowout *n*: an uncontrolled flow of gas, oil, or other well fluids into the atmosphere or into an underground formation. A blowout, or gusher, can occur when formation pressure exceeds the pressure applied to it by the column of drilling fluid.

blowout preventer (BOP) *n*: one of several valves installed at the wellhead to prevent the escape of pressure either in the annular space between the casing and the drill pipe or in open hole (i.e., hole with no drill pipe) during drilling or completion operations. See *annular blowout preventer, ram blowout preventer.*

blowout preventer control panel *n*: controls, usually located near the driller's position on the rig floor, that are manipulated to open and close the blowout preventers. See *blowout preventer.*

blowout preventer control unit *n*: a device that stores hydraulic fluid under pressure in special containers and provides a method to open and close the blowout preventers quickly and reliably. Usually, compressed air and hydraulic pressure provide the opening and closing force in the unit. See *blowout preventer.* Also called an accumulator.

BOP *abbr*: blowout preventer.

BOP stack *n*: the assembly of blowout preventers installed on a well.

bore *n*: the inside diameter of a pipe or a drilled hole. *v*: to penetrate or pierce with a rotary tool.

borehole *n*: a hole made by drilling or boring; a wellbore.

bottomhole *n*: the lowest or deepest part of a well. *adj*: pertaining to the bottom of the wellbore.

bottomhole assembly *n*: the portion of the drilling assembly below the drill pipe. It can be very simple—composed of only the bit and drill collars—or it can be very complex and made up of several drilling tools.

bottomhole pressure *n*: the pressure at the bottom of a borehole caused by the hydrostatic pressure of the wellbore fluid and, sometimes, by any back-pressure held at the surface, as when the well is shut in with blowout preventers.

bottom plug *n*: a cement plug that precedes cement slurry down the casing. The plug wipes drilling mud off the walls of the casing and prevents it from contaminating the cement. See *cementing, wiper plug.*

box *n*: the female section of a connection. See also *tool joint.*

box and pin *n*: see *tool joint.*

box threads *n pl*: threads on the female section, or box, of a tool joint. See *tool joint.*

brake *n*: a device for arresting the motion of a mechanism, usually by means of friction, as in the drawworks brake.

brake band *n*: a part of the brake mechanism consisting of a flexible steel band lined with a material that grips a drum when tightened. On a drilling rig, the brake band acts on the flanges of the drawworks drum to control the lowering of the traveling block and its load of drill pipe, casing, or tubing.

break *v*: to begin or start (e.g., to break circulation or to break tour).

break circulation *v*: to start the mud pump for restoring circulation of the mud column.

break it *v*: see *break out.*

break it out *v*: see *break out.*

break out *v*: to unscrew one section of pipe from another section, especially drill pipe while it is being withdrawn from the wellbore.

breakout block *n*: a bit breaker; a heavy plate that fits in the rotary table and holds the drill bit while it is being unscrewed from the drill collar.

breakout cathead *n*: a device attached to the catshaft of the drawworks that is used as a power source for unscrewing drill pipe; usually located opposite the driller's side of the drawworks. See *cathead.* Compare *makeup cathead.*

breakout tongs *n pl*: tongs that are used to start unscrewing one section of pipe from another section, especially drill pipe coming out of the hole. See *lead tongs, tongs.*

break tour *v*: to begin operating 24 hours a day.

bring in a well *v*: to complete a well and put it on producing status.

buck up *v*: to tighten up a threaded connection (such as two joints of drill pipe).

bulk tank *n*: on a drilling rig, a large metal bin that usually holds a large amount of a certain mud additive, such as bentonite, that is used in large quantities in the makeup of the drilling fluid. Also called a P-tank.

bullwheel *n*: one of the two large wheels joined by an axle and used to hold the drilling line on a cable-tool rig.

bumped *adj*: in cementing operations, pertaining to a cement plug that comes to rest on the float collar. A cementing operator may say, "I have a bumped plug" when the plug strikes the float collar.

bumps *v*: see *bumped.*

bushing *n*: 1. a pipe fitting on which the external thread is larger than the internal thread to allow two pipes of different sizes to be connected. 2. a removable lining or sleeve inserted or screwed into an opening to limit its size, resist wear or corrosion, or serve as a guide.

C

cable *n*: a rope of wire, hemp, or other strong fibers. See *wire rope*.

cable-tool drilling *n*: a drilling method in which the hole is drilled by dropping a sharply pointed bit on bottom. The bit is attached to a cable, and the cable is repeatedly dropped as the hole is drilled.

cable-tool rig *n*: a drilling rig that uses wire rope (cable) to suspend a weighted, chisel-shaped bit in the hole. Machinery on the rig repeatedly lifts and drops the cable and bit. Each time the bit strikes the bottom of the hole, it drills deeper. Rotary drilling rigs have virtually replaced all cable-tool rigs.

caisson *n*: one of several columns made of steel or concrete that serve as the foundation for a rigid offshore platform rig, such as the concrete gravity platform rig.

cap *n*: to control a well that is flowing out of control; often accomplished by attaching a valve in the open position on top of the well and then closing it to seal off the flow

carrier rig *n*: a large, specially designed, self-propelled workover rig that is driven directly to the well site. Power from a carrier rig's hoist engine or engines also propels the rig on the road. While a carrier rig is primarily intended to perform workovers, it can also be used to drill relatively shallow wells. A carrier rig may be a back-in type or a drive-in type. Compare *back-in unit, drive-in unit..*

case *n*: the outer cylinder of a concentric cylinder centrifuge. See *concentric cylinder centrifuge.*

cased *adj*: pertaining to a wellbore in which casing has been run and cemented. See *casing.*

cased hole *n*: a wellbore in which casing has been run.

casing *n*: steel pipe placed in an oil or gas well to prevent the wall of the hole from caving in, to prevent movement of fluids from one formation to another, and to improve the efficiency of extracting petroleum if the well is productive.

casing centralizer *n*: a device secured around the casing at regular intervals to center it in the hole. Casing that is centralized allows a more uniform cement sheath to form around the pipe.

casing coupling *n*: a tubular section of pipe that is threaded inside and used to connect two joints of casing.

casing crew *n*: the employees of a company that specializes in preparing and running casing into a well. Usually, the casing crew makes up the casing as it is lowered into the well; however, the regular drilling crew also assists the casing crew in its work.

casing elevator *n*: see *elevators.*

casing float collar *n*: see *float collar.*

casing float shoe *n*: see *float shoe.*

casing hanger *n*: a circular device with a frictional gripping arrangement of slips and packing rings used to suspend casing from a casinghead in a well.

casinghead *n*: a heavy, flanged steel fitting connected to the first string of casing. It provides a housing for slips and packing assemblies, allows suspension of intermediate and production strings of casing, and supplies the means for the annulus to be sealed off. Also called a spool.

casing point *n*: 1. the depth in a well at which casing is set, generally the depth at which the casing shoe rests. 2. the objective depth in a drilling contract, either a specified depth or the depth at which a specific zone is penetrated.

casing pressure *n*: the pressure in a well that exists between the casing and the drill pipe.

casing seat *n*: the location of the bottom of a string of casing that is cemented in a well.

casing shoe *n*: see *guide shoe.*

casing string *n*: the entire length of all the joints of casing run in a well.

casing tongs *n pl*: large wrench used for turning when making up or breaking out casing. See *tongs.*

casing-tubing annulus *n*: in a wellbore, the space between the inside of the casing and the outside of the tubing.

catch samples *v*: to obtain cuttings for geological information as formations are penetrated by the bit.

cathead *n*: a spool-shaped attachment on the end of the catshaft, around which rope for hoisting and moving heavy equipment on or near the rig floor is wound. See *breakout cathead, makeup cathead.*

cathead spool *n*: see *cathead.*

catline *n*: a hoisting or pulling line powered by the cathead and used to lift heavy equipment on the rig. See *cathead.*

catshaft *n*: an axle that crosses through the drawworks and contains a revolving spool called a cathead at either end. See *cathead.*

catwalk *n*: the ramp at the side of the drilling rig where pipe is laid to be lifted to the derrick floor by the catline or by an air hoist.

caving *n*: collapsing of the walls of the wellbore. Also called sloughing.

cellar *n*: a pit in the ground, usually lined with concrete or wood, that provides additional height between the rig floor and the wellhead to accommodate the installation of blowout preventers, rathole, mousehole, and so forth.

cement *n*: a powder consisting of alumina, silica, lime, and other substances that hardens when mixed with water. Extensively used in the oil industry to bond casing to the walls of the wellbore.

cement casing *v*: to fill the annulus between the casing and the wall of the hole with cement to support the casing and to prevent fluid migration between permeable zones.

cementing *n*: the application of a liquid slurry of cement and water to various points inside or outside the casing.

cementing company *n*: a company whose specialty is preparing, transporting, and pumping cement into a well. Usually, a cementing company's crew pumps the cement to secure casing in the well.

cementing head *n*: an accessory attached to the top of the casing to facilitate cementing of the casing. It has passages for cement slurry and retainer chambers for cementing wiper plugs. Also called retainer head.

cementing pump *n*: a high-pressure pump used to force cement down the casing and into the annular space between the casing and the wall of the borehole.

cementing time *n*: the total elapsed time needed to complete a cementing operation.

cement plug *n*: 1. a portion of cement placed at some point in the wellbore to seal it. 2. a wiper plug. See *cementing, wiper plug.*

centralizer *n*: see *casing centralizer.*

chain tongs *n pl*: a hand tool consisting of a handle and chain that resembles the chain on a bicycle. In general, chain tongs are used for turning pipe or fittings of a diameter larger than that which a pipe wrench would fit.

choke *n*: a device with an orifice installed in a line to restrict the flow of fluids.

choke line *n*: a pipe attached to the blowout preventer stack out of which kick fluids and mud can be pumped to the choke manifold when a blowout preventer is closed in on a kick.

choke manifold *n*: an arrangement of piping and special valves, called chokes. In drilling, mud is circulated through a choke manifold when the blowout preventers are closed.

Christmas tree *n*: the control valves, pressure gauges, and chokes assembled at the top of a well to control the flow of oil and gas after the well has been drilled and completed.

circulate *v*: to pass from one point throughout a system and back to the starting point. For example, drilling fluid is circulated out of the suction pit, down the drill pipe and drill collars, out the bit, up the annulus, and back to the pits while drilling proceeds.

circulating components *n pl*: the equipment included in the drilling fluid circulating system of a rotary rig. Basically, the components consist of the mud pump, the rotary hose, the swivel, the drill stem, the bit, and the mud return line.

circulating fluid *n*: see *drilling fluid, mud.*

circulating pressure *n*: the pressure generated by the mud pumps and exerted on the drill stem.

circulation *n*: the movement of drilling fluid out of the mud pits, down the drill stem, up the annulus, and back to the mud pits.

clay *n*: 1. a term used for particles smaller than 1/256 millimetre (4 microns), regardless of mineral composition. 2. a group of hydrous aluminum silicate minerals (clay minerals).

clump weights *n pl*: special segmented weights attached to the guy wires of a guyed compliant platform that keep the guy wires taut as the platform jacket moves with the waves and current of the water.

coiled tubing *n*: a continuous string of flexible steel tubing, often hundreds or thousands of feet long, that is wound onto a reel, often dozens of feet in diameter. The reel is an integral part of the coiled tubing unit, which consists of several devices that ensure the tubing can be safely and efficiently inserted into the well from the surface. Because tubing can be lowered into a well without having to make up joints of tubing, running coiled tubing into the well is faster and less expensive than running conventional tubing. Rapid advances in the use of coiled tubing make it a popular way in which to run tubing into and out of a well. Also called reeled tubing.

collapse pressure *n*: the amount of force needed to crush the sides of pipe until it caves in.

collar *n*: 1. a coupling device used to join two lengths of pipe, such as casing or tubing. 2. a drill collar. See *drill collar.*

collar locator *n*: a logging device used to determine accurately the depth of a well; the log measures and records the depth of each casing collar, or coupling, in a well.

combination trap *n*: 1. a subsurface hydrocarbon trap that has the features of both a structural trap and a stratigraphic trap. 2. a combination of two or more structural traps or two or more stratigraphic traps.

come out of the hole *v*: to pull the drill stem out of the wellbore to change the bit, to change from a core barrel to the bit, to run logs, to prepare for a drill stem test, to run casing, and so on. Also called trip out.

commercial quantity *n*: an amount of oil and gas production large enough to enable the operator to realize a profit, however small.

compact *n*: a small disk made of tungsten carbide. See *insert*.

company hand *n*: see *company representative*.

company man *n*: see *company representative*.

company representative *n*: an employee of an operating company whose job is to represent the company's interests at the drilling location.

complete a well *v*: to finish work on a well and bring it to productive status. See *well completion*.

compliant platform *n*: an offshore platform that is designed to flex with wind and waves.

compound *n*: a mechanism used to transmit power from the engines to the pump, the drawworks, and other machinery on a drilling rig. It is composed of clutches, chains and sprockets, belts and pulleys, and a number of shafts, both driven and driving. *v*: to connect two or more power-producing devices, such as engines, to run driven equipment, such as the drawworks.

compresion-ignition engine *n*: a diesel engine; an engine in which the fuel-air mixture inside the engine cylinders is ignited by the heat that occurs when the fuel-air mixture is highly compressed by the engine pistons.

concrete gravity platform rig *n*: a rigid offshore drilling platform built of steel-reinforced concrete and used to drill development wells. See *platform rig*.

conductor casing *n*: generally, the first string of casing in a well. Its purpose is to prevent the soft formations near the surface from caving in and to conduct drilling mud from the bottom of the hole to the surface when drilling starts. Also called conductor pipe, drive pipe.

conductor pipe *n*: see *conductor casing*.

cone *n*: a conical-shaped metal device into which cutting teeth are formed or mounted on a roller cone bit. See *roller cone bit*.

confirmation well *n*: the second producer in a new field, following the discovery well.

connection *n*: 1. the action of adding a joint of pipe to the drill stem as drilling progresses. 2. a section of pipe or a fitting used to join pipe to pipe or to a vessel.

contract *n*: a written agreement that can be enforced by law and that lists the terms under which the acts required are to be performed. A drilling contract covers such factors as the cost of drilling the well (whether by the foot or by the day), the distribution of expenses between operator and contractor, and the type of equipment to be used.

contract depth *n*: the depth of the wellbore at which a drilling contract is fulfilled.

controlled directional drilling *n*: see *directional drilling*.

core *n*: a cylindrical sample taken from a formation for geological analysis. *v*: to obtain a solid, cylindrical formation sample for analysis.

core barrel *n*: a tubular device, usually from 10 to 60 feet (3 to 18 metres) long, run at the bottom of the drill pipe in place of a bit and used to cut a core sample.

crane *n*: a machine for raising, lowering, and revolving heavy pieces of equipment, especially on offshore rigs and platforms.

crew *n*: the workers on a drilling rig, including the driller, the derrickhand, and the rotary helpers.

crossover sub *n*: a sub that allows different sizes and types of drill pipe to be joined.

crown *n*: the crown block or top of a derrick or mast.

crown block *n*: an assembly of sheaves mounted on beams at the top of the derrick or mast and over which the drilling line is reeved. See *block*.

crude oil *n*: unrefined liquid petroleum. It ranges in density from very light to very heavy and in color from yellow to black, and it may have a paraffin, asphalt, or mixed base.

cutters *n pl*: 1. cutting teeth on the cones of a roller cone bit. 2. the parts of a reamer that actually contact the wall of the hole and open it to full gauge. A three-point reamer has three cutters; a six-point reamer has six cutters. Cutters are available for different formations.

cuttings *n pl*: the fragments of rock dislodged by the bit and brought to the surface in the drilling fluid.

D

daily drilling report *n*: a record made each day of the operations on a working drilling rig and, traditionally, phoned or radioed in to the office of the drilling company every morning. Also called morning report.

daylight tour *n*: in areas where three 8-hour tours are worked, the shift on a drilling rig that starts at or about daylight. Compare *afternoon tour* and *morning tour*.

day tour *n*: in areas where two 12-hour tours are worked, a period of 12 daylight hours worked by a drilling crew.

daywork *adj*: descriptive of work done on daywork rates.

daywork rates *n pl*: the basis for payment on drilling contracts when the rig owner is paid by the day rather than by the foot. Daywork rates are the most common way in which contractors are paid for the rig's work.

deadline *n*: the drilling line from the crown block sheave to the anchor, so called because it does not move. Compare *fastline*.

deadline anchor *n*: see *deadline tie-down anchor*.

deadline tie-down anchor *n*: a device to which the deadline is attached, securely fastened to the mast or derrick substructure. Also called a deadline anchor.

deflect *n*: see *deflection*.

deflection *n*: a change in the angle of a wellbore. In directional drilling, it is measured in degrees from the vertical.

degasser *n*: the device used to remove gas from drilling fluid.

density *n*: the mass or weight of a substance per unit volume.

derrick *n*: a large load-bearing structure, usually of bolted construction. In drilling, the standard derrick has four legs standing at the corners of the substructure and reaching to the crown block. Compare *mast*.

derrick floor *n*: also called the rig floor or the drill floor. See *rig floor*.

derrickhand *n*: the crew member who handles the upper end of the drill string as it is being hoisted out of or lowered into the hole. This person is also responsible for the circulating machinery and the conditioning of the drilling fluid.

derrickman *n*: see *derrickhand*.

desander *n*: a centrifugal device for removing sand from drilling fluid to prevent abrasion of the pumps. Compare *desilter*.

desilter *n*: a centrifugal device for removing very fine particles, or silt, from drilling fluid to keep the amount of solids in the fluid at the lowest possible point. Compare *desander*.

development well *n*: 1. a well drilled in proven territory in a field to complete a pattern of production. 2. an exploitation well.

diamond bit *n*: a drill bit that has small industrial diamonds embedded in its cutting surface. Cutting is performed by the rotation of the very hard diamonds over the rock surface.

diapir *n*: a dome or anticlinal fold in which a mobile plastic core has ruptured the more brittle overlying rock. Also called piercement dome.

dies *n pl*: tools used to shape, form, or finish other tools or pieces of metal. For example, a threading die is used to cut threads on pipe.

diesel-electric rig *n*: see *electric rig*.

diesel engine *n*: a high-compression, internal-combustion engine used extensively for powering drilling rigs. In a diesel engine, air is drawn into the cylinders and compressed to very high pressures; ignition occurs as fuel is injected into the compressed and heated air. Combustion takes place within the cylinder above the piston, and expansion of the combustion products imparts power to the piston.

directional drilling *n*: intentional deviation of a wellbore from the vertical.

directional hole *n*: a wellbore intentionally drilled at an angle from the vertical.

discovery well *n*: the first oil or gas well drilled in a new field that reveals the presence of a hydrocarbon-bearing reservoir.

displacement fluid *n*: in oilwell cementing, the fluid, usually drilling mud or salt water, that is pumped into the well after the cement is pumped into it to force the cement out of the casing and into the annulus.

doghouse *n*: a small enclosure on the rig floor used as an office for the driller and as a storehouse for small objects.

double *n*: a length of drill pipe, casing, or tubing consisting of two joints screwed together. Compare *fourble*, *single*, *thribble*.

double board *n*: the name used for the derrickhand's working platform (the monkeyboard) when it is located at a height in the derrick or mast equal to two lengths of pipe joined together. Compare *fourble board*, *thribble board*.

downhole *adj*, *adv*: pertaining to the wellbore.

downhole motor *n*: a drilling tool made up in the drill string directly above the bit. It causes the bit to turn while the drill string remains fixed.

Drake well *n*: the first well drilled in the United States in search of oil. Some 69 feet (21 metres) deep, it was drilled near Titusville, Pennsylvania, and was completed in 1859. It was named after Edwin L. Drake, who was hired by the well owners to oversee the drilling.

drawworks *n*: the hoisting mechanism on a drilling rig. It is essentially a large winch that spools off or takes in the drilling line and thus raises or lowers the drill stem and the bit.

drawworks brake *n*: the mechanical brake on the drawworks that can prevent the drawworks drum from moving.

drawworks drum *n*: the spool-shaped cylinder in the drawworks around which drilling line is wound, or spooled.

drill *v*: to bore a hole in the earth, usually to find and remove subsurface formation fluids such as oil and gas.

drill ahead *v*: to continue drilling operations.

drill bit *n*: the cutting or boring element used for drilling. See *bit*.

drill collar *n*: a heavy, thick-walled tube, usually steel, placed between the drill pipe and the bit in the drill stem.

drill collar sub *n*: a sub made up between the drill string and the drill collars that is used to ensure that the drill pipe and the collar can be joined properly.

drill column *n*: see *drill stem*.

drilled show *n*: oil or gas in the mud circulated to the surface.

driller *n*: the employee directly in charge of a drilling rig and crew. This person's main duty is operation of the drilling and hoisting equipment, but the driller is also responsible for downhole condition of the well, operation of downhole tools, and pipe measurements.

driller's console *n*: a metal cabinet on the rig floor containing the controls that the driller manipulates to operate various components of the drilling rig.

driller's control panel *n*: see *driller's console*.

driller's position *n*: the area immediately surrounding the driller's console.

drill floor *n*: also called rig floor or derrick floor. See *rig floor*.

drilling contract *n*: an agreement made between a drilling company and an operating company to drill and complete a well. It sets forth the obligation of each party, compensation, identification, method of drilling, depth to be drilled, and so on.

drilling contractor *n*: an individual or group that owns a drilling rig or rigs and contracts services for drilling wells.

drilling crew *n*: a driller, a derrickhand, and two or more rotary helpers who operate a drilling rig.

drilling engine *n*: an internal-combustion engine used to power a drilling rig.

drilling engineer *n*: an engineer who specializes in the technical aspects of drilling.

drilling fluid *n*: circulating fluid, one function of which is to lift cuttings out of the wellbore and to the surface. Other functions are to cool the bit and to counteract downhole formation pressure. Although a mixture of clay and other minerals, water, and chemical additives is the most common drilling fluid, wells can also be drilled by using air, gas, water, or oil-base mud as the drilling mud. See *mud*.

drilling hook *n*: the large hook mounted on the bottom of the traveling block and from which the swivel is suspended. When drilling, the entire weight of the drill stem is suspended from the hook.

drilling line *n*: a wire rope used to support the drilling tools. Also called the rotary line.

drilling mud *n*: a specially compounded liquid circulated through the wellbore during rotary drilling operations. See *drilling fluid, mud*.

drilling rate *n*: the speed with which the bit drills the formation; usually called the rate of penetration (ROP).

drill pipe *n*: seamless steel or aluminum pipe made up in the drill stem between the kelly or top drive on the surface and the drill collars on the bottom. Several joints are made up (screwed together) to form the drill string.

drill pipe slips *n pl*: see *slips*.

drill ship *n*: a self-propelled floating offshore drilling unit that is a ship constructed to permit a well to be drilled from it. Although not as stable as semisubmersibles, drill ships are capable of drilling exploratory wells in deep, remote waters. See *floating offshore drilling rig*.

drill site *n*: the location of a drilling rig.

drill stem *n*: all members in the assembly used for rotary drilling from the swivel to the bit, including the kelly, the drill pipe and tool joints, the drill collars, the stabilizers, and various specialty items. Compare *drill string*.

drill stem test (DST) *n*: the conventional method of formation testing. The basic drill stem test tool consists of a packer or packers, valves or ports that may be opened and closed from the surface, and two or more pressure-recording devices. The tool is lowered on the drill string to the zone to be tested. The packer or packers are set to isolate the zone from the drilling fluid column. The valves or ports are then opened to allow for formation flow while the recorders chart static pressures. A sampling chamber traps clean formation fluids at the end of the test.

drill string *n*: the column, or string, of drill pipe with attached tool joints that transmits fluid and rotational power from the kelly to the drill collars and the bit. Often, especially in the oil patch, the term is loosely applied to both drill pipe and drill collars. Compare *drill stem*.

drive bushing *n*: see *kelly bushing.*

drive-in unit *n*: a type of portable service or workover rig that is self-propelled, using power from the hoisting engines. The driver's cab and steering wheel are mounted on the same end as the mast support; thus the unit can be driven straight ahead to reach the wellhead. See *carrier rig.*

drive pipe *n*: see *conductor casing.*

drum *n*: a cylinder around which wire rope is wound in the drawworks. The drawworks drum is that part of the hoist on which the drilling line is wound.

dry *n*: a hole is dry when the reservoir it penetrates is not capable of producing hydrocarbons in commercial amounts.

dry hole *n*: any well that does not produce oil or gas in commercial quantities.

dynamic positioning *n*: a method by which a floating offshore drilling rig is maintained in position over an offshore well location without the use of mooring anchors. Generally, several propulsion units, called thrusters, are located on the hulls of the structure and are actuated by a sensing system. A computer to which the system feeds signals directs the thrusters to maintain the rig on location.

dynamic positioning operator *n*: an employee on a drill ship or semisubmersible drilling rig whose primary duty is to monitor, operate, and maintain the equipment that maintains the rig on station while drilling.

E

electric drive *n*: see *electric rig.*

electric-drive rig *n*: see *electric rig.*

electric generator *n*: a machine that changes mechanical energy into electrical energy.

electric rig *n*: a drilling rig on which the energy from the power source—usually several diesel engines—is changed to electricity by generators mounted on the engines. Compare *mechanical rig.*

elevator bails *n pl*: see *elevator links.*

elevator links *n pl*: cylindrical bars that support the elevators and attach them to the hook. Also called elevator bails.

elevators *n pl*: clamps that grip a joint of casing, tubing, drill collars, or drill pipe so that the joint can be raised from or lowered into the hole.

engine *n*: a machine for converting the heat content of fuel into rotary motion that can be used to power other machines.

evening tour *n*: see *afternoon tour.*

exploitation well *n*: a well drilled to permit more effective extraction of oil from a reservoir. Sometimes called a development well.

exploration *n*: the search for reservoirs of oil and gas, including aerial and geophysical surveys, geological studies, core testing, and drilling of wildcats.

exploration well *n*: a well drilled either in search of an as-yet-undiscovered pool of oil or gas (a wildcat well) or to extend greatly the limits of a known pool.

F

fastline *n*: the end of the drilling line that is affixed to the drum or reel of the drawworks, so called because it travels with greater velocity than any other portion of the line. Compare *deadline.*

fault *n*: a break in the earth's crust along which rocks on one side have been displaced (upward, downward, or laterally) relative to those on the other side.

fault trap *n*: a subsurface hydrocarbon trap created by faulting, in which an impermeable rock layer has moved opposite the reservoir bed or where impermeable gouge has sealed the fault and stopped fluid migration.

female connection *n*: a pipe, a coupling, or a tool threaded on the inside so that only a male connection can be joined to it. Compare *male connection.*

field *n*: a geographical area in which a number of oil or gas wells produce from a continuous reservoir. A field may refer to surface area only or to underground productive formations as well.

fingerboard *n*: a rack that supports the tops of the stands of pipe being stacked in the derrick or mast.

fish *n*: an object that is left in the wellbore during drilling or workover operations and that must be recovered before work can proceed. *v*: to recover from a well any equipment left there during drilling operations, such as a lost bit or drill collar or part of the drill string.

fishing *n*: the procedure of recovering lost or stuck equipment in the wellbore. See also *fish.*

fishing tool *n*: a tool designed to recover equipment lost in a well.

fixed-head bit *n*: any bit, such as a diamond bit, whose cutting elements do not move on the face, or head, of the bit. Compare *roller cone bit.*

flexible drill pipe *n*: specially manufactured drill pipe that has several pressure-tight joints over the length of the pipe. These joints allow the pipe to bend considerably more than regular drill pipe and are used in directional wells (especially horizontal ones) where the angle of deflection from vertical is relatively abrupt.

flex joint *n*: a device that provides a flexible connection between the riser pipe and the subsea blowout preventers.

float collar *n*: a special coupling device inserted one or two joints above the bottom of the casing string that contains a check valve to permit fluid to pass downward but not upward through the casing. The float collar prevents drilling mud from entering the casing while it is being lowered, allowing the casing to float during its descent and thus decreasing the load on the derrick or mast. A float collar also prevents backflow of cement during a cementing operation.

floater *n*: see *floating offshore drilling rig*.

floating offshore drilling rig *n*: a type of mobile offshore drilling unit that floats and is not in contact with the seafloor (except with anchors) when it is in the drilling mode. Floating units include drill ships and semisubmersibles. See *mobile offshore drilling unit.*

float shoe *n*: a short, heavy, cylindrical steel section with a rounded bottom that is attached to the bottom of the casing string. It contains a check valve and functions similarly to the float collar but also serves as a guide shoe for the casing.

floe *n*: a floating ice field of any size.

floor crew *n*: those workers on a drilling or workover rig who work primarily on the rig floor.

floorhand *n*: see *rotary helper*.

floorman *n*: see *rotary helper*.

fluid *n*: a substance that flows and yields to any force tending to change its shape. Liquids and gases are fluids.

footage rates *n pl*: a fee basis in drilling contracts stipulating that payment to the drilling contractor is made according to the number of feet or metres of hole drilled.

forge *n*: to use hard blows to form and shape metallic ingots into useful items.

formation *n*: a bed or deposit composed throughout of substantially the same kind of rock. Each formation is given a name, frequently as a result of the study of the formation outcrop at the surface and sometimes based on fossils found in the formation.

formation boundary *n*: the horizontal limits of a formation.

formation fluid *n*: fluid (such as gas, oil, or water) that exists in a subsurface rock formation.

formation fracturing *n*: a method of stimulating production by opening new flow channels in the rock surrounding a production well. Often called a frac job. Under extremely high hydraulic pressure, a fluid (such as distillate, diesel fuel, crude oil, dilute hydrochloric acid, water, or kerosene) is pumped downward through production tubing or drill pipe and forced out below a packer or between two packers. The pressure causes cracks to open in the formation, and the fluid penetrates the formation through the cracks. Sand grains, aluminum pellets, walnut shells, or similar materials (propping agents) are carried in suspension by the fluid into the cracks. When the pressure is released at the surface, the fracturing fluid returns to the well. The cracks partially close on the pellets, leaving channels for oil to flow around them to the well.

formation pressure *n*: the force exerted by fluids in a formation, recorded in the hole at the level of the formation with the well shut in. Also called reservoir pressure or shut-in bottomhole pressure.

fourble *n*: a section of drill pipe, drill collars, or tubing consisting of four joints screwed together. Compare *double, single, thribble.*

fourble board *n*: the name used for the derrickhand's working platform, or the monkeyboard, when it is located at a height in the derrick equal to approximately four lengths of pipe joined together. Compare *double board, thribble board.*

frac job *n*: see *formation fracturing.*

fracturing *n*: shortened form of formation fracturing. See *formation fracturing.*

full-gauge bit *n*: a bit that has maintained its original diameter.

full-gauge hole *n*: a wellbore drilled with a full-gauge bit. Also called a true-to-gauge hole.

G

gas-cut mud *n*: a drilling mud that contains entrained formation gas, giving the mud a characteristically fluffy texture. When entrained gas is not released before the fluid returns to the well, the weight or density of the fluid column is reduced.

gas drilling *n*: see *air drilling.*

gauge *n*: 1. the diameter of a bit or the hole drilled by the bit. 2. a device (such as a pressure gauge) used to measure some physical property. *v*: to measure size, volume, depth, or other measurable property.

gel *n*: a semisolid, jellylike state assumed by some colloidal dispersions at rest. When agitated, the gel converts to a fluid state. Also a nickname for bentonite. *v*: to take the form of a gel; to set.

geologist *n*: a scientist who gathers and interprets data pertaining to the rocks of the earth's crust.

geology *n*: the science of the physical history of the earth and its life, especially as recorded in the rocks of the crust.

geophone *n*: an instrument placed on the surface that detects vibrations passing through the earth's crust. It is used in conjunction with seismography.

geophysicist *n*: one who studies geophysics.

Geronimo *n*: see *safety slide*.

go in the hole *v*: to lower the drill stem, the tubing, or the casing into the wellbore.

gooseneck *n*: the curved connection between the rotary hose and the swivel. See *swivel*.

grief stem *n*: (obsolete) kelly; kelly joint.

guidelines *n pl*: lines, usually four, attached to a special guide base to help position equipment (such as blowout preventers) accurately on the seafloor when a well is drilled offshore from a floating vessel.

guide shoe *n*: a short, heavy, cylindrical section of steel filled with concrete and rounded at the bottom, which is placed at the end of the casing string. A passage through the center of the shoe allows drilling fluid to pass up into the casing while it is being lowered and allows cement to pass out during cementing operations. Also called casing shoe.

gusher *n*: an oilwell that has come in with such great pressure that the oil jets out of the well like a geyser. In reality, a gusher is a blowout and is extremely wasteful of reservoir fluids and drive energy. See *blowout*.

guyed-tower platform rig *n*: a compliant offshore drilling platform used to drill development wells. The foundation of the platform is a relatively lightweight jacket on which all equipment is placed. A system of guy wires anchored by clump weights helps secure the jacket to the seafloor and allows it to move with wind and wave forces. See *platform rig*.

H

head *n*: 1. the height of a column of liquid required to produce a specific pressure. See *hydraulic head*. 2. for centrifugal pumps, the velocity of flowing fluid converted into pressure expressed in feet or metres of flowing fluid. Also called velocity head. 3. that part of a machine (such as a pump or an engine) that is on the end of the cylinder opposite the crankshaft. Also called cylinder head.

hex kelly *n*: see *kelly*.

hoist *n*: 1. an arrangement of pulleys and wire rope or chain used for lifting heavy objects; a winch or similar device. 2. the drawworks. *v*: to raise or lift.

hoisting components *n pl*: drawworks, drilling line, and traveling and crown blocks. Auxiliary hoisting components include catheads, catshaft, and air hoist.

hoisting drum *n*: the large flanged spool in the drawworks on which the hoisting cable is wound. See *drawworks*.

hole *n*: in drilling operations, the wellbore or borehole. See *borehole*, *wellbore*.

hook *n*: a large, hook-shaped device from which the swivel is suspended. It is designed to carry maximum loads ranging from 100 to 650 tons (90 to 590 tonnes) and turns on bearings in its supporting housing.

hook load *n*: the weight of the drill stem that is suspended from the hook.

hopper *n*: a large funnel- or cone-shaped device into which dry components (such as powdered clay or cement) can be poured to mix uniformly with water or other liquids.

horizontal drilling *n*: deviation of the borehole at least 80° from vertical so that the borehole penetrates a productive formation in a manner parallel to the formation.

horsepower *n*: a unit of measure of work done by a machine. One horsepower equals 33,000 foot-pounds per minute. (Kilowatts are used to measure power in the international, or SI, system of measurement.)

hybrid bits *n pl*: combine natural and synthetic diamonds and sometimes tungsten carbide inserts on a fixed-head bit.

hydraulic *adj*: 1. of or relating to water or other liquid in motion. 2. operated, moved, or affected by water or liquid.

hydraulic fracturing *n*: an operation in which a specially blended liquid is pumped down a well and into a formation under pressure high enough to cause the formation to crack open, forming passages through which oil can flow into the wellbore.

hydraulic head *n*: the force exerted by a column of liquid expressed by the height of the liquid above the point at which the pressure is measured. Although "head" refers to distance or height, it is used to express pressure, since the force of the liquid column is directly proportional to its height. Also called head or hydrostatic head. Compare *hydrostatic pressure*.

hydrocarbons *n pl*: organic compounds of hydrogen and carbon whose densities, boiling points, and freezing points increase as their molecular weights increase. Although composed of only two elements, hydrocarbons exist in a variety of compounds, because of the strong affinity of the carbon atom for other atoms and for itself. Petroleum is a mixture of many different hydrocarbons.

hydrogas *n*: another term for liquefied petroleum gas (LPG).

hydrophone *n*: a device trailed in an array behind a boat in offshore seismic exploration that is used to detect sound reflections, convert them to electric current, and send them through a cable to recording equipment on the boat.

hydrostatic pressure *n*: the force exerted by a body of fluid at rest. It increases directly with the density and the depth of the fluid and is expressed in pounds per square inch or kilopascals. The hydrostatic pressure of fresh water is 0.433 pounds per square inch per foot (9.792 kilopascals per metre) of depth. In drilling, the term refers to the pressure exerted by the drilling fluid in the wellbore. In a water drive field, the term refers to the pressure that may furnish the primary energy for production.

I

IADC *abbr*: International Association of Drilling Contractors.

idiot stick *n*: (slang) a shovel.

ignorant end *n*: (slang) the heavier end of any device (such as a length of pipe or a wrench).

independent *n*: a nonintegrated oil company or an individual whose operations are in the field of petroleum production, excluding transportation, refining, and marketing.

infill drilling *n*: drilling wells between known producing wells to exploit the resources of a field to best advantage.

infilling well *n*: a well drilled between known producing wells to exploit the reservoir better.

inland barge rig *n*: a floating offshore drilling structure consisting of a barge on which the drilling equipment is constructed. When moved from one location to another, the barge floats. When stationed on the drill site, the barge can be anchored in the floating mode or submerged to rest on the bottom. Typically, inland barge rigs are used to drill wells in marshes, shallow inland bays, and areas where the water is not too deep. Also called swamp barge. See *floating offshore drilling rig*.

insert *n*: a cylindrical object, rounded, blunt, or chisel-shaped on one end and usually made of tungsten carbide, that is inserted in the cones of a bit, the cutters of a reamer, or the blades of a stabilizer to form the cutting element of the bit or the reamer or the wear surface of the stabilizer. Also called a compact.

intermediate casing string *n*: the string of casing set in a well after the surface casing but before production casing is set to keep the hole from caving and to seal off troublesome formations. In deep wells, one or more intermediate strings may be required. Sometimes called protection casing.

intermediate string *n*: see *intermediate casing string*.

internal-combustion engine *n*: a heat engine in which the pressure necessary to produce motion of the mechanism results from the ignition or burning of a fuel-air mixture within the engine cylinder.

International Association of Drilling Contractors (IADC) *n*: an organization of drilling contractors that sponsors or conducts research on education, accident prevention, drilling technology, and other matters of interest to drilling contractors and their employees. Its official publication is *The Drilling Contractor*. Address: Box 4287; Houston, TX 77210; (713) 578-7171.

international system of units (SI) *n*: a system of units of measurement based on the metric system, adopted and described by the Eleventh General Conference on Weights and Measures. It provides an international standard of measurement to be followed when certain customary units, both U.S. and metric, are eventually phased out of international trade operations.

Iron Roughneck™ *n*: a manufacturer's name for a floor-mounted combination of a spinning wrench and a torque wrench. The Iron Roughneck™ moves into position hydraulically and eliminates the manual handling involved with suspended individual tools.

J

jackup drilling rig *n*: a mobile bottom-supported offshore drilling structure with columnar or open-truss legs that support the deck and hull. When positioned over the drilling site, the bottoms of the legs rest on the seafloor. A jackup rig is towed or propelled to a location with its legs up. Once the legs are firmly positioned on the bottom, the deck and the hull height are adjusted and leveled. Also called self-elevating drilling unit.

jerk line *n*: a length of chain used on the automatic cathead of the drilling rig to tighten pipe joints by pulling on the makeup tongs. See *makeup tongs*.

jet *n*: 1. a hydraulic device operated by a centrifugal pump used to clean the mud pits, or tanks, and to mix mud components. 2. in a perforating gun using shaped charges, a highly penetrating, fast-moving stream of exploded particles that forms a hole in the casing, cement, and formation.

jet bit *n*: a drilling bit having replaceable nozzles through which the drilling fluid is directed in a high-velocity stream to the bottom of the hole to improve the efficiency of the bit. See *bit*.

jet-perforate *v*: to create holes through the casing with a shaped charge of high explosives.

joint *n*: a single length (from 16 feet to 45 feet, or 5 metres to 14.5 metres, depending on its range length) of drill pipe, drill collar, casing, or tubing that has threaded connections at both ends.

joint of pipe *n*: a length of drill pipe or casing. Both come in various lengths. See *range length*.

joule *n*: the SI unit of energy or work. It is equal to 1 newton-metre (n•m), which is 1 newton of force acting through a distance of 1 metre in the direction of the force.

junk *n*: metal debris lost in a hole.

K

kelly *n*: the heavy steel tubular device, four- or six-sided, suspended from the swivel through the rotary table and connected to the top joint of drill pipe to turn the drill stem as the rotary table turns. It has a bored passageway that permits fluid to be circulated into the drill stem and up the annulus, or vice versa.

kelly bushing *n*: a special device placed around the kelly that mates with the kelly flats and fits into the master bushing of the rotary table. Also called the drive bushing.

kelly cock *n*: a valve installed at one or both ends of the kelly that is closed when a high-pressure backflow begins inside the drill stem. The valve is closed to keep pressure off the swivel and rotary hose.

kelly drive bushing *n*: a device that fits into the master bushing of the rotary table and through which the kelly runs. When the master bushing rotates the kelly drive bushing, the kelly drive bushing rotates the kelly and the drill stem attached to the kelly.

kelly flat *n*: one of the flat sides of a kelly. Also called a flat.

kelly hose *n*: also called the mud hose or rotary hose. See *rotary hose*.

kelly joint *n*: see *kelly*.

kelly saver sub *n*: a sub that fits in the drill stem between the kelly and the drill pipe and prevents wear to the kelly's threads.

kelly spinner *n*: a pneumatically operated device mounted on top of the kelly that, when actuated, causes the kelly to turn, or spin. It is used when making up or breaking out the kelly from the drill string.

kelly sub *n*: see *kelly saver sub*.

kick *n*: an entry of water, gas, oil, or other formation fluid into the wellbore during drilling. It occurs because the pressure exerted by the column of drilling fluid is not great enough to overcome the pressure exerted by the fluids in the formation drilled.

kick fluids *n pl*: oil, gas, water, or any combination that enters the borehole from a permeable formation.

kick off *v*: to deviate a wellbore from the vertical, as in directional drilling.

kickoff point (KOP) *n*: the depth in a vertical hole at which a deviated or slant hole is started; used in directional drilling.

kill *v*: to control a kick by taking suitable preventive measures (e.g., to shut in the well with the blowout preventers, circulate the kick out, and increase the weight of the drilling mud).

L

land rig *n*: any drilling rig that is located on dry land. Compare *offshore rig*.

latch on *v*: to attach elevators to a section of pipe to pull it out of or run it into the hole.

lead-tong hand *n*: the crew member who operates the lead tongs when drill pipe and drill collars are being handled.

lead tongs *n pl*: the pipe tongs suspended in the derrick or mast and operated by a chain or a wire rope connected to the makeup cathead or the breakout cathead. Personnel call the makeup tongs the lead tongs if pipe is going into the hole; similarly, they call the breakout tongs the lead tongs if pipe is coming out of the hole.

lens *n*: 1. a porous, permeable, irregularly shaped sedimentary deposit surrounded by impervious rock. 2. a lenticular sedimentary bed that pinches out, or comes to an end, in all directions.

lens-type trap *n*: a hydrocarbon reservoir consisting of a porous, permeable, irregularly shaped sedimentary deposit surrounded by impervious rock. See *lens*.

lifting nipple *n*: also called hoisting plug or lifting sub. See *lifting sub*.

lifting sub *n*: a short piece of pipe with a pronounced upset, or shoulder, on the upper end, screwed into drill collars to provide a positive grip for the elevators. Also called a lifting nipple or a hoisting plug.

liner *n*: a string of pipe used to case open hole below existing casing. A liner extends from the setting depth up into another string of casing, usually overlapping about 100 feet (30 metres) into the upper string.

liner hanger *n*: a slip device that attaches the liner to the casing. See *liner*.

liquid *n*: a state of matter in which the shape of the given mass depends on the containing vessel, but the volume of the mass is independent of the vessel. A liquid is a fluid that is almost incompressible.

location *n*: the place where a well is drilled. Also called well site.

log *n*: a systematic recording of data, such as a driller's log, a mud log, an electrical well log, or a nuclear log. *v*: to record data.

log a well *v*: to run any of the various logs used to ascertain downhole information about a well.

logging devices *n pl*: any of several electrical, acoustical, mechanical, or nuclear devices that are used to measure and record certain characteristics or events that occur in a well that has been or is being drilled.

logging while drilling (LWD) *n*: logging measurements obtained by measurement-while-drilling techniques as the well is being drilled.

lost time accident *n*: an incident in the work place that results in an injury serious enough that causes the person injured to be unable to work for a day or more.

M

major *n*: a large oil company, such as ExxonMobil or Chevron, that not only produces oil, but also transports, refines, and markets it and its products.

make a connection *v*: to attach a joint of drill pipe onto the drill stem suspended in the wellbore to permit deepening the wellbore by the length of the joint (usually about 30 feet, or 9 metres).

make a trip *v*: to hoist the drill stem out of the wellbore to perform one of a number of operations, such as changing bits or taking a core, and then to return the drill stem to the wellbore.

make hole *v*: to deepen the hole made by the bit, i.e., to drill ahead.

make up *v*: 1. to assemble and join parts to form a complete unit (e.g., to make up a string of casing). 2. to screw together two threaded pieces. 3. to mix or prepare (e.g., to make up a tank of mud).

makeup *adj*: added to a system (e.g., makeup water used in mixing mud).

make up a joint *v*: to screw a length of pipe into another length of pipe.

makeup cathead *n*: a device that is attached to the shaft of the drawworks and used as a power source for screwing together joints of pipe. It is usually located on the driller's side of the drawworks. Also called spinning cathead. See *cathead*. Compare *breakout cathead*.

makeup tongs *n pl*: tongs used for screwing one length of pipe into another for making up a joint. See *lead tongs*, *tongs*.

male connection *n*: a pipe, a coupling, or a tool that has threads on the outside so that it can be joined to a female connection. Compare *female connection*.

manifold *n*: an accessory system of piping to a main piping system (or another conductor) that serves to divide a flow into several parts, to combine several flows into one, or to reroute a flow to any one of several possible destinations.

marine riser connector *n*: a fitting on top of the subsea blowout preventers to which the riser pipe is connected.

marine riser pipe *n*: see *riser pipe*.

marine riser system *n*: see *riser pipe*.

mast *n*: a portable derrick that is capable of being raised as a unit. Compare *derrick*.

master bushing *n*: a device that fits into the rotary table to accommodate the slips and drive the kelly bushing so that the rotating motion of the rotary table can be transmitted to the kelly. Also called rotary bushing.

measurement while drilling (MWD) *n*: 1. directional and other surveying during routine drilling operations to determine the angle and direction by which the wellbore deviates from the vertical. 2. any system of measuring downhole conditions during routine drilling operations.

mechanical-drive rig *n*: see *mechanical rig*.

mechanical rig *n*: a drilling rig in which the source of power is one or more internal-combustion engines and in which the power is distributed to rig components through mechanical devices (such as chains, sprockets, clutches, and shafts). Also called a power rig. Compare *electric rig*.

megajoule (MJ) *n*: the SI unit of service given by a drilling line when it moves 1,000 newtons of load over a distance of 1,000 metres.

metre (m) *n*: the fundamental unit of length in the international system of measurement (SI). It is equal to about 3.28 feet, 39.37 inches, or 100 centimetres.

metric system *n*: a decimal system of weights and measures based on the metre as the unit of length, the gram as the unit of weight, the cubic metre as the unit of volume, the litre as the unit of capacity, and the square metre as the unit of area. The international system of measurement (SI) is based on the metric system.

metric ton *n*: a measurement equal to 1,000 kilograms or 2,204.6 avoirdupois. In some oil-producing countries, production is reported in metric tons. One metric ton is equivalent to about 7.4 barrels (42 U.S. gallons = 1 barrel) of crude oil. In the SI system it is called a tonne.

mill *n*: a downhole tool with rough, sharp, extremely hard cutting surfaces for removing metal by grinding or cutting. They are also called junk mills, reaming mills, and so forth, depending on their use. *v*: to use a mill to cut or grind metal objects that must be removed from a well.

mineral rights *n pl*: the rights of ownership, conveyed by deed, of gas, oil, and other minerals beneath the surface of the earth. In the United States, mineral rights are the property of the surface owner unless disposed of separately.

mix mud *v*: to prepare drilling fluids from a mixture of water or other liquids and any one or more of the various dry mud-making materials (such as clay, weighting materials, and chemicals).

MJ *abbr*: megajoule.

mobile offshore drilling unit (MODU) *n*: a drilling rig that is used to drill offshore exploration and development wells. It floats on the surface of the water when being moved from one drill site to another, but it may or may not float once drilling begins.

MODU *abbr*: mobile offshore drilling unit.

monkeyboard *n*: the derrickhand's working platform. As pipe is run into or out of the hole, the derrickhand must handle the top end of the pipe, which may be 90 feet (27 metres) or higher in the derrick or mast. The monkeyboard provides a small platform to raise the derrickhand to the proper elevation for handling the top of the pipe.

morning tour *n*: on rigs in which crews work three 8-hour tours, the tour that typically starts around midnight and lasts until 7:00 or 8:00 A.M.

motorhand *n*: the crew member on a rotary drilling rig, usually the most experienced rotary helper, who is responsible for the care and operation of drilling engines.

motorman *n*: see *motorhand*.

mousehole *n*: an opening through the rig floor, usually lined with pipe, into which a length of drill pipe is placed temporarily for later connection to the drill string.

mousehole connection *n*: the procedure of adding a length of drill pipe to the active string. The length to be added is placed in the mousehole, made up to the kelly, then pulled out of the mousehole and subsequently made up into the string. Compare *rathole connection*.

mud *n*: the liquid circulated through the wellbore during rotary drilling operations. In addition to its function of bringing cuttings to the surface, drilling mud cools and lubricates the bit and the drill stem, protects against blowouts by holding back subsurface pressures, and deposits a mud cake on the wall of the borehole to prevent loss of fluids to the formation. See *drilling fluid*.

mud centrifuge *n*: a device that uses centrifugal force to separate small solid components from liquid drilling fluid.

mud cleaner *n*: a cone-shaped device, a hydrocyclone, designed to remove very fine solid particles from the drilling mud.

mud engineer *n*: an employee of a drilling fluid supply company whose duty it is to test and maintain the drilling mud properties that are specified by the operator.

mud-gas separator *n*: a device that removes gas from the mud coming out of a well when a kick is being circulated out.

mud hopper *n*: see *hopper*.

mud hose *n*: also called kelly hose or rotary hose. See *rotary hose*.

mud line *n*: 1. in offshore operations, the seafloor. 2. a mud return line.

mud logger *n*: an employee of a mud logging company who performs mud logging.

mud logging *n*: the recording of information derived from examination and analysis of formation cuttings made by the bit and of mud circulated out of the hole.

mud pit *n*: originally, an open pit dug in the ground to hold drilling fluid or waste materials discarded after the treatment of drilling mud. For some drilling operations, mud pits are used for suction to the mud pumps, settling of mud sediments, and storage of reserve mud. Steel tanks are much more commonly used for these purposes now, but they are still referred to as pits, except offshore, where "mud tanks" is preferred.

mud pump *n*: a large, high-pressure reciprocating pump used to circulate the mud on a drilling rig. Also called a slush pump.

mud return line *n*: a trough or pipe that is placed between the surface connections at the wellbore and the shale shaker and through which drilling mud flows on its return to the surface from the hole. Also called flow line.

mud tank *n*: one of a series of open tanks, usually made of steel plate, through which the drilling mud is cycled to remove sand and fine sediments. Also called mud pits.

mud weight *n*: a measure of the density of a drilling fluid expressed as pounds per gallon, pounds per cubic foot, or kilograms per cubic metre. Mud weight is directly related to the amount of pressure the column of drilling mud exerts at the bottom of the hole.

N

natural gas *n*: a highly compressible, highly expandable mixture of hydrocarbons with a low specific gravity and occurring naturally in a gaseous form. Besides hydrocarbon gases, natural gas may contain appreciable quantities of nitrogen, helium, carbon dioxide, hydrogen sulfide, and water vapor.

newton (N) *n*: an SI unit that expresses force. One newton equals 1 metre-kilogram per second per second ($m \cdot kg/s^2$), which is the force required to move 1 kilogram a distance of 1 metre at a velocity of 1 second squared.

night toolpusher *n*: an assistant toolpusher whose duty hours are typically during nighttime hours on a mobile offshore drilling unit.

nipple up *v*: to assemble the blowout preventer stack on the wellhead at the surface.

nonporous *adj*: containing no interstices; having no pores.

normal circulation *n*: the smooth, uninterrupted circulation of drilling fluid down the drill stem, out the bit, up the annular space between the pipe and the hole, and back to the surface. Compare *reverse circulation*.

nozzle *n*: a passageway through jet bits that causes the drilling fluid to be ejected from the bit at high velocity. The jets of mud clear the bottom of the hole.

O

offset well *n*: a well drilled in the vicinity of other wells to assess the extent and characteristics of the reservoir and, in some cases, to drain hydrocarbons from an adjoining lease or tract.

offshore *n*: that geographic area that lies seaward of the coastline. In general, the term "coastline" means the line of ordinary low water along that portion of the coast that is in direct contact with the open sea or the line marking the seaward limit of inland waters.

offshore drilling *n*: drilling for oil or gas in an ocean, gulf, or sea, usually on the Outer Continental Shelf. A drilling unit for offshore operations may be a mobile floating vessel with a ship or barge hull, a semisubmersible or submersible base, a self-propelled or towed structure with jacking legs (jackup drilling rig), or a permanent structure used as a production platform when drilling is completed.

offshore installation manager (OIM) *n*: a qualified and certified person with marine and drilling knowledge who is in charge of all operations on a MODU.

offshore production platform *n*: an immobile offshore structure from which wells are produced.

offshore rig *n*: any of various types of drilling structures designed for use in drilling wells in oceans, seas, bays, gulfs, and so forth. Offshore rigs include platforms, jackup drilling rigs, semisubmersible drilling rigs, and drill ships. Compare *land rig*.

oil *n*: a simple or complex liquid mixture of hydrocarbons that can be refined to yield gasoline, kerosene, diesel fuel, and various other products.

oil-base mud *n*: a drilling fluid in which oil is the continuous phase and which contains from less than 2 percent and up to 5 percent water.

oilfield *n*: the surface area overlying an oil reservoir or reservoirs. The term usually includes not only the surface area, but also the reservoir, the wells, and the production equipment.

oil mud *n*: a drilling mud, e.g., oil-base mud and invert-emulsion mud, in which oil is the continuous phase.

oil patch *n*: (slang) the oilfield.

oil sand *n*: 1. a sandstone that yields oil. 2. (by extension) any reservoir that yields oil, whether or not it is sandstone.

oil seep *n*: a surface location where oil appears, the oil having permeated its subsurface boundaries and accumulated in small pools or rivulets. Also called oil spring.

oilwell *n*: a well from which oil is obtained.

oilwell cement *n*: cement or a mixture of cement and other materials for use in oil, gas, or water wells.

oil zone *n*: a formation or horizon of a well from which oil may be produced. The oil zone is usually immediately under the gas zone and on top of the water zone if all three fluids are present and segregated.

open *adj*: 1. of a wellbore, having no casing. 2. of a hole, having no drill pipe or tubing suspended in it.

open hole *n*: 1. any wellbore in which casing has not been set. 2. open or cased hole in which no drill pipe or tubing is suspended. 3. the portion of the wellbore that has no casing.

open-hole fishing *n*: the procedure of recovering lost or stuck equipment in an uncased wellbore.

operating company *n*: see *operator*.

operator *n*: the person or company actually operating an oilwell, generally the oil company that hires a drilling contractor.

organic theory *n*: an explanation of the origin of petroleum that holds that the hydrogen and the carbon that make up petroleum come from land and sea plants and animals. The theory further holds that more of this organic material comes from very tiny swamp and sea creatures than comes from larger land creatures.

Organization of Petroleum Exporting Countries (OPEC) *n*: an organization of the countries of the Middle East, Southeast Asia, Africa, and South America that produce oil and export it. Members as of 1997 are Algeria, Ecuador, Gabon, Indonesia, Iran, Iraq, Kuwait, Libya, Nigeria, Qatar, Saudi Arabia, the United Arab Emirates, and Venezuela. The organization's purpose is to negotiate and regulate oil prices.

orientation *n*: the process of positioning a deflection tool so that it faces in the direction necessary to achieve the desired direction and drift angle for a directional hole.

Outer Continental Shelf (OCS) *n*: the land seaward from areas subject to state mineral ownership to a depth of roughly 8,000 feet (2,500 metres), beyond which mineral exploration and development are not, at present, feasible. Boundaries of the OCS are set by law. In general, the term is used to describe federally controlled areas.

out-of-gauge bit *n*: a bit that is no longer of the proper diameter.

out-of-gauge hole *n*: a hole that is not in gauge; that is, it is smaller or larger than the diameter of the bit used to drill it.

outpost well *n*: a well located outside the established limits of a reservoir, i.e., a step-out well.

overshot *n*: a fishing tool that is attached to tubing or drill pipe and lowered over the outside wall of pipe lost or stuck in the wellbore. A friction device in the overshot, usually either a basket or a spiral grapple, firmly grips the pipe, allowing the fish to be pulled from the hole.

P

P&A *abbr*: plug and abandon.

packer *n*: a piece of downhole equipment that consists of a sealing device, a holding or setting device, and an inside passage for fluids. It is used to block the flow of fluids through the annular space between pipe and the wall of the wellbore by sealing off the space between them. A packing element expands to prevent fluid flow except through the packer and tubing.

pay *n*: see *pay sand*.

pay formation *n*: see *pay sand*.

pay sand *n*: the producing formation, often one that is not sandstone. Also called pay, pay zone, and producing zone.

pay zone *n*: see *pay sand*.

PDC *abbr*: polycrystalline diamond compact.

PDC bit *n*: a special type of diamond drilling bit that does not use roller cones. Instead, polycrystalline diamond inserts, or compacts, are embedded into a matrix on the bit.

penetration rate *n*: see *rate of penetration*.

perforate *v*: to pierce the casing wall and cement of a wellbore to provide holes through which formation fluids may enter or to provide holes in the casing so that materials may be introduced into the annulus between the casing and the wall of the borehole.

perforated completion *n*: 1. a well completion method in which the producing zone or zones are cased through, cemented, and perforated to allow fluid flow into the wellbore. 2. a well completed by this method.

perforated liner *n*: a liner that has had holes shot in it by a perforating gun. See *liner*.

perforating gun *n*: a device fitted with shaped charges or bullets that is lowered to the desired depth in a well and fired to create penetrating holes in casing, cement, and formation.

perforating truck *n*: a special vehicle designed to allow control of a perforating operation within it.

perforation *n*: a hole made in the casing, cement, and formation through which formation fluids enter a wellbore. Usually several perforations are made at a time.

permeability *n*: 1. a measure of the ease with which a fluid flows through the connecting pore spaces of a rock. The unit of measurement is the millidarcy. 2. fluid conductivity of a porous medium. 3. ability of a fluid to flow within the interconnected pore network of a porous medium.

permeable *adj*: allowing the passage of fluid. See *permeability*.

petroleum *n*: a substance occurring naturally in the earth in solid, liquid, or gaseous state and composed mainly of mixtures of chemical compounds of carbon and hydrogen, with or without other nonmetallic elements such as sulfur, oxygen, and nitrogen. In some cases, especially in the measurement of oil and gas, petroleum refers only to oil—a liquid hydrocarbon—and does not include natural gas or gas liquids such as propane and butane. The API prefers that petroleum mean crude oil and not natural gas or gas liquids.

petroleum geology *n*: the study of oil- and gas-bearing rock formations. It deals with the origin, the occurrence, the movement, and the accumulation of hydrocarbon fuels.

pick up *v*: 1. to use the drawworks to lift the bit (or other tool) off bottom by raising the drill stem. 2. to use an air hoist to lift a tool, a joint of drill pipe, or other piece of equipment.

piercement dome *n*: see *diapir*.

pin *n*: 1. the male section of a tool joint. 2. on a bit, the bit shank, which screws into a bit sub or drill collar.

pinch-out *n*: an oil-bearing stratum that forms a trap for oil and gas by narrowing and tapering off within an impervious formation.

pinch out *v*: to end or terminate by a narrowing and tapering off. When a formation pinches out, it narows and tapers off.

pipe *n*: a long, hollow cylinder, usually steel, through which fluids are conducted.

pipe rack *n*: a horizontal support for tubular goods.

pipe racker *n*: a pneumatic or hydraulic device that, on command from an operator, either picks up pipe from a rack or from the side of the derrick and lifts it into the derrick or takes pipe from out of the derrick and places it on the rack or places it to the side of the derrick.

pipe ram *n*: a sealing component for a blowout preventer that closes the annular space between the pipe and the blowout preventer or wellhead.

pipe ram preventer *n*: a blowout preventer that uses pipe rams as the closing elements. See *pipe ram*.

pipe tongs *n pl*: see *tongs*.

pipe upset *n*: that part of the pipe that has an abrupt increase of dimension.

pit level *n*: height of drilling mud in the mud tanks, or pits.

platform *n*: see *platform rig*.

platform jacket *n*: a support that is firmly secured to the ocean floor and to which the legs of a platform are anchored.

platform rig *n*: an immobile offshore structure from which development wells are drilled and produced. Platform rigs may be built of steel or concrete and may be rigid or compliant. Rigid platform rigs, which rest on the seafloor, are the concrete gravity platform and the steel-jacket platform. Compliant platform rigs, which are used in deeper waters and yield to water and wind movements, are the guyed-tower platform and the tension-leg platform.

play *n*: 1. the extent of a petroleum-bearing formation. 2. the activities associated with petroleum development in an area.

plug and abandon (P&A) *v*: to place cement plugs into a dry hole and abandon it.

plug container *n*: see *cementing head*.

pneumatic *adj*: operated by air pressure.

polycrystalline diamond compact (PDC) *n*: a synthetic diamond used in the manufacture of the cutters on PDC bits.

pontoon *n*: an attachment, added to a stinger, that is flooded to lower pipeline toward the seafloor at an angle that will not overstress it.

porosity *n*: 1. the condition of being porous (such as a rock formation). 2. the ratio of the volume of empty space to the volume of solid rock in a formation, indicating how much fluid a rock can hold.

porous *adj*: having pores, or tiny openings, as in rock.

pore *n*: an opening or space within a rock or mass of rocks, usually small and often filled with some fluid (water, oil, gas, or all three). Compare *vug*.

possum belly *n*: 1. a receiving tank situated at the end of the mud return line. The flow of mud comes into the bottom of the device and travels over baffles to control mud flow over the shale shaker. 2. a metal box under a truck bed that holds repair tools.

power rig *n*: see *mechanical rig*.

power swivel *n*: a top drive. See *top drive*.

pressure *n*: the force that a fluid (liquid or gas) exerts uniformly in all directions within a vessel, a pipe, a hole in the ground, and so forth, such as that exerted against the inner wall of a tank or that exerted on the bottom of the wellbore by a fluid. Pressure is expressed in terms of force exerted per unit of area, as pounds per square inch, or in kilopascals.

preventer *n*: shortened form of blowout preventer. See *blowout preventer*.

prime mover *n*: an internal-combustion engine or a turbine that is the source of power for driving a machine or machines.

producer *n*: 1. a well that produces oil or gas in commercial quantities. 2. an operating company or individual in the business of producing oil; commonly called the operator.

producing horizon *n*: see *pay sand*.

producing interval *n*: see *pay sand*.

producing platform *n*: an offshore structure accommodating a number of producing wells.

producing zone *n*: the zone or formation from which oil or gas is produced. See *pay sand*.

production *n*: 1. the phase of the petroleum industry that deals with bringing the well fluids to the surface and separating them and with storing, gauging, and otherwise preparing the product for the pipeline. 2. the amount of oil or gas produced in a given period.

production casing *n*: the last string of casing set in a well, the inside of which is usually suspended a tubing string.

production platform *n*: see *platform rig*.

production well *n*: the well through which oil is produced in fields where improved recovery techniques are being applied.

proppant *n*: see *propping agent*.

propping agent *n*: a granular substance (sand grains, aluminum pellets, or other material) that is carried in suspension by the fracturing fluid and that serves to keep the cracks open when fracturing fluid is withdrawn after a fracture treatment.

P-tank *n*: see *bulk tank*.

pulley *n*: a wheel with a grooved rim, used for pulling or hoisting.

pull out *v*: see *come out of the hole*.

pull singles *v*: to remove the drill stem from the hole by disconnecting each individual joint.

pump *n*: a device that increases the pressure on a fluid or raises it to a higher level.

pup joint *n*: a length of drill or line pipe, tubing, or casing shorter than range 1 (18 feet or 6.26 metres for drill pipe) in length.

pusher *n*: shortened form of toolpusher.

Q

quadruples *n*: see *fourbles*.

R

rack *n*: framework for supporting or containing a number of loose objects, such as pipe. *v*: 1. to place on a rack. 2. to use as a rack.

ram *n*: the closing and sealing component on a blowout preventer. One of three types—blind, pipe, or shear—may be installed in several preventers mounted in a stack on top of the wellbore.

ram blowout preventer *n*: a blowout preventer that uses rams to seal off pressure on a hole that is with or without pipe. Also called ram preventer. Compare *annular blowout preventer*.

ram preventer *n*: see *ram blowout preventer*.

range length *n*: a grouping of pipe lengths. API designation of range lengths is as follows:

Range 1		Range 2		Range 3	
(feet)	(metres)	(feet)	(metres)	(feet)	(metres)
CASING					
16–25	4.88–7.62	25–34	7.62–10.36	34–48	10.36–14.63
DRILL PIPE					
18–22	5.49–6.71	27–30	8.23–9.14	38–45	11.58–13.72
TUBING					
20–24	6.10–7.32	28–32	8.53–9.75		

rate of penetration (ROP) *n*: a measure of the speed at which the bit drills into formations, usually expressed in feet (metres) per hour or minutes per foot (metre).

rathole *n*: a hole in the rig floor, which is lined with casing that projects above the floor and into which the kelly and swivel are placed when hoisting operations are in progress.

rathole connection *n*: the addition of a length of drill pipe to the active string using the rathole instead of the mousehole, which is the more common connection. Compare *mousehole connection*.

rathole rig *n*: a small, usually truck-mounted rig, the purpose of which is to drill ratholes for regular drilling rigs that will be moved in later. A rathole rig may also drill the top part of the hole, the conductor hole, before the main rig arrives on location.

reamer *n*: a tool used in drilling to smooth the wall of a well, enlarge the hole to the specified size, help stabilize the bit, straighten the wellbore if kinks or doglegs are encountered, and drill directionally.

reciprocation *n*: a back-and-forth or up-and-down movement (as the movement of a piston in an engine or pump).

reel *n*: a revolving device (such as a flanged cylinder) for winding or unwinding something flexible (such as rope or wire).

reeve *v*: to pass (as a rope) through a hole or opening in a block or similar device.

reeve the line *v*: to string a wire rope drilling line through the sheaves of the traveling and the crown blocks to the hoisting drum.

remote BOP control panel *n*: a device placed on the rig floor that can be operated by the driller to direct air pressure to actuating cylinders that turn the control valves on the main BOP control unit, located a safe distance from the rig.

remote choke panel *n*: a set of controls, usually placed on the rig floor, that is manipulated to control the amount of drilling fluid being circulated through the choke manifold. See *choke manifold*.

reserve pit *n*: a waste pit, usually an excavated earthen-walled pit. It may be lined with plastic or other material to prevent soil contamination.

reservoir *n*: a subsurface, porous, permeable rock body in which oil and/or gas has accumulated.

reservoir pressure *n*: the average pressure within the reservoir at any given time.

reservoir rock *n*: a permeable rock that may contain oil or gas in appreciable quantity and through which petroleum may migrate.

retainer head *n*: see *cementing head*.

reverse circulation *n*: the course of drilling fluid downward through the annulus and upward through the drill stem. Also referred to as "circulating the short way," since returns from bottom can be obtained more quickly than in normal circulation. Compare *normal circulation*.

rig *n*: the derrick or mast, drawworks, and attendant surface equipment of a drilling unit.

rig crew member *n*: see *rotary helper*.

rig down *v*: to dismantle a drilling rig and auxiliary equipment following the completion of drilling operations. Also called tear down.

rig floor *n*: the area immediately around the rotary table and extending to each corner of the derrick or mast—that is, the area immediately above the substructure on which the drawworks, the rotary table, and so forth rest. Also called derrick floor, drill floor.

rig manager *n*: an employee of a drilling contractor who is in charge of the entire drilling crew and the drilling rig, providing logistics support to the rig crew and liaison with the operating company.

rig superintendent *n*: see *toolpusher*.

rig supervisor *n*: see *toolpusher*.

rig up *v*: to prepare the drilling rig for making hole, i.e., to install tools and machinery before drilling is started.

riser pipe *n*: the pipe and special fittings used on floating offshore drilling rigs to establish a seal between the top of the wellbore, which is on the ocean floor, and the drilling equipment, located above the surface of the water. A riser pipe serves as a guide for the drill stem from the drilling vessel to the wellhead and as a conductor of drilling fluid from the well to the vessel. Also called marine riser.

riser tensioner line *n*: a cable that supports the marine riser while compensating for vessel movement.

rock *n*: a hardened aggregate of different minerals. Rocks are divided into three groups on the basis of their mode of origin: igneous, metamorphic, and sedimentary.

rock bit *n*: see *roller cone bit*.

rock oil *n*: see *petroleum*.

roller bit *n*: see *roller cone bit*.

roller chain *n*: a type of chain that is used to transmit power by fitting over sprockets attached to shafts, causing rotation of one shaft by the rotation of another.

roller cone bit *n*: a drilling bit made of two, three, or four cones that are mounted on extremely rugged bearings. The surface of each cone has rows of steel teeth or rows of tungsten carbide inserts. Also called rock bits.

rotary *n*: the machine used to impart rotational power to the drill stem while permitting vertical movement of the pipe for rotary drilling.

rotary bushing *n*: see *master bushing*.

rotary drilling *n*: a drilling method in which a hole is drilled by a rotating bit to which a downward force is applied. The bit is fastened to and rotated by the drill stem, which also provides a passageway through which the drilling fluid is circulated.

rotary drilling rig *n*: a drilling rig that features a system that rotates a bit and, at the same time, has a system that continuously circulates drilling fluid while drilling is going on.

rotary helper *n*: a worker on a drilling or workover rig, subordinate to the driller, whose primary work station is on the rig floor. Sometimes called floorhand, floorman, rig crewman, or roughneck.

rotary hose *n*: a reinforced flexible tube on a rotary drilling rig that conducts the drilling fluid from the standpipe to the swivel and kelly. Also called the mud hose or the kelly hose.

rotary speed *n*: the speed, measured in revolutions per minute, at which the rotary table is operated.

rotary support table *n*: a strong but relatively lightweight device used on some rigs that employ a top drive to rotate the bit. Although a conventional rotary table is not required to rotate the bit on such rigs, crew members must still have a place to set the slips to suspend the drill string in the hole when tripping or making a connection. A rotary support table provides such a place but does not include all the rotary machinery required in a regular rotary table.

rotary table *n*: the principal component of a rotary, or rotary machine, used to turn the drill stem and support the drilling assembly.

rotary-table system *n*: a series of devices that provide a way to rotate the drill stem and bit. Basic components consist of a turntable, master bushing, kelly drive bushing, kelly, and a swivel.

rotary torque *n*: the rotational force applied to turn the drill stem.

rotate on bottom *v*: see *make hole*.

rotating components *n pl*: those parts of the drilling or workover rig that are designed to turn or rotate the drill stem and bit—swivel, kelly, kelly bushing, master bushing, and rotary table.

roughneck *n*: see *rotary helper*.

round trip *n*: the action of pulling out and subsequently running back into the hole a string of drill pipe or tubing. Making a round trip is also called tripping.

roustabout *n*: a worker on an offshore rig who handles the equipment and supplies that are sent to the rig from the shore base. The head roustabout is very often the crane operator.

run casing *v*: to lower a string of casing into the hole. Also called to run pipe.

run in *v*: to go into the hole with tubing, drill pipe, and so forth.

run pipe *v*: to lower a string of casing into the hole. Also called to run casing.

S

safety clamp *n*: a clamp placed very tightly around a drill collar that is suspended in the rotary table by drill collar slips. Should the slips fail, the clamp is too large to go through the opening in the rotary table and therefore prevents the drill collar string from falling into the hole.

safety slide *n*: a wireline device normally mounted near the monkeyboard to afford the derrickhand a means of quick exit to the surface in case of emergency. It is usually affixed to a wireline, one end of which is attached to the derrick or mast and the other end to the surface. To exit by the safety slide, the derrickhand grasps a handle on it and rides it down to the ground. Also called a Geronimo.

salt dome *n*: a dome that is caused by an intrusion of rock salt into overlying sediments. A piercement salt dome is one that has been pushed up so that it penetrates the overlying sediments, leaving them truncated.

samples *n pl*: 1. the well cuttings obtained at designated footage intervals during drilling. From an examination of these cuttings, the geologist determines the type of rock and formations being drilled and estimates oil and gas content. 2. small quantities of well fluids obtained for analysis.

sand *n*: 1. an abrasive material composed of small quartz grains formed from the disintegration of preexisting rocks. Sand consists of particles less than 0.062 inch (2 millimetres) and greater than 0.078 inch (//¡§ millimetre) in diameter. 2. sandstone.

sandstone *n*: a sedimentary rock composed of individual mineral grains of rock fragments between 0.062 and 0.078 inch (//¡§ and 2 millimetres) in diameter and cemented together by silica, calcite, iron oxide, and so forth. Sandstone is commonly porous and permeable and therefore a likely type of rock in which to find a petroleum reservoir.

saver sub *n*: a device made up in the drill stem to absorb much of the wear between frequently broken joints (such as between the kelly and the drill pipe). See *kelly saver sub*.

scratcher *n*: a device that is fastened to the outside of casing to remove mud cake from the wall of a hole to condition the hole for cementing. By rotating or moving the casing string up and down as it is being run into the hole, the scratcher, formed of stiff wire, removes the cake so that the cement can bond solidly to the formation.

seafloor *n*: the bottom of the ocean; the seabed.

sediment *n*: in geology, buried layers of sedimentary rocks.

sedimentary rock *n*: a rock composed of materials that were transported to their present position by wind or water. Sandstone, shale, and limestone are sedimentary rocks.

seat *n*: the point in the wellbore at which the bottom of the casing is set.

seep *n*: the surface appearance of oil or gas that results naturally when a reservoir rock becomes exposed to the surface, thus allowing oil or gas to flow out of fissures in the rock.

seismic *adj*: of or relating to an earthquake or earth vibration, including those artificially induced.

seismic data *n*: detailed information obtained from earth vibration produced naturally or artificially (as in geophysical prospecting).

seismic method *n*: a method of geophysical prospecting using the generation, reflection, refraction detection, and analysis of sound waves in the earth.

seismic survey *n*: an exploration method in which strong low-frequency sound waves are generated on the surface or in the water to find subsurface rock structures that may contain hydrocarbons. Interpretation of the record can reveal possible hydrocarbon-bearing formations.

self-elevating substructure *n*: a base on which the floor and mast of a drilling rig rests and which, after it is placed in the desired location, is raised into position as a single unit.

self-propelled unit *n*: see *carrier rig*.

semisubmerged *n*:a state in which a specially designed floating drilling rig (a semisubmersible) floats just below the water's surface.

semisubmersible *n*: see *semisubmersible drilling rig*.

semisubmersible drilling rig *n*: a floating offshore drilling unit that has pontoons and columns that, when flooded, cause the unit to submerge to a predetermined depth. Semisubmersibles are more stable than drill ships and are used extensively to drill wildcat wells in rough waters such as the North Sea. See *floating offshore drilling rig*.

set back *v*: to place stands of drill pipe and drill collars in a vertical position to one side of the rotary table in the derrick or mast of a drilling or workover rig.

set casing *v*: to run and cement casing at a certain depth in the wellbore. Sometimes called set pipe.

set pipe *v*: see *set casing*.

shaker *n*: shortened form of shale shaker. See *shale shaker*.

shale *n*: a fine-grained sedimentary rock composed mostly of consolidated clay or mud. Shale is the most frequently occurring sedimentary rock.

shale shaker *n*: a vibrating screen used to remove cuttings from the circulating fluid in rotary drilling operations. The size of the openings in the screen should be carefully selected to be the smallest size possible to allow 100 percent flow of the fluid. Also called a shaker.

shaped charge *n*: a relatively small container of high explosive that is loaded into a perforating gun. On detonation, the charge releases a small, high-velocity stream of particles (a jet) that penetrates the casing, cement, and formation. See *perforating gun.*

shear ram *n*: the component in a blowout preventer that cuts, or shears, through drill pipe and forms a seal against well pressure.

shear ram preventer *n*: a blowout preventer that uses shear rams as closing elements.

sheave *n*: a grooved pulley.

shoulder *n*: the flat portion of a tool joint.

shut in *v*: 1. to close the valves on a well so that it stops producing. 2. to close in a well in which a kick has occurred.

single *n*: a joint of drill pipe. Compare *double, fourble,* and *thribble.*

sinker bar *n*: a heavy weight or bar placed on or near a lightweight wireline tool. The bar provides weight so that the tool will lower properly into the well.

slack off *v*: to lower a load or ease up on a line. A driller will slack off on the brake to put additional weight on the bit.

slingshot substructure *n*: see *self-elevating substructure*.

slip and cutoff program *n*: a procedure to ensure that the drilling line wears evenly throughout its life. After a specified number of ton-miles (megajoules) of use, the line is slipped—i.e., the traveling block is suspended in the derrick or propped on the rig floor so that it cannot move, the deadline anchor bolts are loosened, and the drilling line is spooled onto the drawworks drum. Enough line is slipped to change the major points of wear on the line, such as where it passes through the sheaves. To prevent excess line from accumulating on the drawworks drum, the worn line is cut off and discarded.

slips *n pl*: wedge-shaped pieces of metal with teeth or other gripping elements that are used to prevent pipe from slipping down into the hole or to hold pipe in place.

sloughing *n*: see *caving.*

sloughing hole *n*: a condition wherein shale that has absorbed water from the drilling fluid expands, sloughs off, and falls downhole. A sloughing hole can jam the drill string and block circulation.

slurry *n*: in drilling, a plastic mixture of cement and water that is pumped into a well to harden. There it supports the casing and provides a seal in the wellbore to prevent migration of underground fluids.

sour crude *n*: see *sour crude oil.*

sour crude oil *n*: oil containing hydrogen sulfide or another acid gas.

spark ignition (SI) *n*: ignition of a fuel-air mixture by means of a spark discharged by a spark plug.

spark-ignition engine *n*: an internal combustion engine that uses an electrical spark to ignite the fuel-air mixture inside its cylinders. Usually, the engine employs spark plugs to provide the electrical spark.

spear *n*: a fishing tool used to retrieve pipe lost in a well. The spear is lowered down the hole and into the lost pipe. When weight, torque, or both are applied to the string to which the spear is attached, the slips in the spear expand and tightly grip the inside of the wall of the lost pipe.

spinning cathead *n*: see *makeup cathead, spinning chain.*

spinning chain *n*: a Y-shaped chain used to spin up (tighten) one joint of drill pipe into another.

spinning wrench *n*: air-powered or hydraulically powered wrench used to spin drill pipe when making up or breaking out connections.

spool *n*: the drawworks drum. Also a casinghead or drilling spool. *v*: to wind around a drum.

spud *v*: to begin drilling a well—i.e., to spud in.

spud in *v*: to begin drilling; to start the hole.

stab *v*: to guide the end of a pipe into a coupling or tool joint when making up a connection.

stabilizer *n*: a tool placed on a drill collar near the bit that is used, depending on where it is placed, either to maintain a particular hole angle or to change the angle by controlling the location of the contact point between the hole and the collars.

stack a rig *v*: to store a drilling rig on completion of a job when the rig is to be withdrawn from operation for a time.

stand *n*: the connected joints of pipe racked in the derrick or mast when making a trip. On a rig, the usual stand is about 90 feet (about 27 metres) long (three lengths of drill pipe screwed together), or a thribble.

standard derrick *n*: a derrick that is built piece by piece at the drilling location. Compare *mast.*

standpipe *n*: a vertical pipe rising along the side of the derrick or mast, which joins the discharge line leading from the mud pump to the rotary hose and through which mud is pumped into the hole.

steel cone *n*: see *roller cone bit.*

steel-jacket platform rig *n*: a rigid offshore drilling platform used to drill development wells. The foundation of the platform is the jacket, a tall vertical section made of tubular steel members. The jacket, which is usually supported by piles driven into the seabed, extends upward so that the top rises above the waterline. Additional sections that provide space for crew quarters, the drilling rig, and all equipment needed to drill are placed on top of the jacket. See *platform rig*.

steel-tooth bit *n*: a roller cone bit in which the surface of each cone is made up of rows of steel teeth. Also called a milled-tooth bit or milled bit.

stem *n*: see *sinker bar*, *swivel stem*.

step-out well *n*: a well drilled adjacent to or near a proven well to ascertain the limits of the reservoir; an outpost well.

stimulation *n*: any process undertaken to enlarge old channels or to create new ones in the producing formation of a well (e.g., acidizing or formation fracturing).

straight hole *n*: a hole that is drilled vertically. The total hole angle is restricted, and the hole does not change direction rapidly—no more than 3° per 100 feet (30.48 metres) of hole.

stratigraphic trap *n*: a petroleum trap that occurs when the top of the reservoir bed is terminated by other beds or by a change of porosity or permeability within the reservoir itself. Compare *structural trap*.

stratum *n*: singular of strata. A distinct, generally parallel bed of rock.

string up *v*: to thread the drilling line through the sheaves of the crown block and the traveling block. One end of the line is secured to the hoisting drum and the other to the derrick substructure.

structural trap *n*: a petroleum trap that is formed because of deformation (such as folding or faulting) of the reservoir formation. Compare *stratigraphic trap*.

structure *n*: a geological formation of interest to drillers.

stuck pipe *n*: drill pipe, drill collars, casing, or tubing that has inadvertently become immovable in the hole.

sub *n*: a short, threaded piece of pipe used to adapt parts of the drilling string that cannot otherwise be screwed together because of differences in thread size or design.

submerged *n*: a state in which a rig that floats on the surface while being moved is in contact with the seafloor when it is in the drilling mode.

subsea blowout preventer *n*: a blowout preventer placed on the seafloor for use by a floating offshore drilling rig.

subsea engineer *n*: see *subsea equipment supervisor*.

subsea equipment supervisor *n*: an employee on a floating offshore drilling rig whose main responsibility is running, monitoring, and maintaining such subsea equipment as the blowout preventer stack, the marine riser system, and similar subsea equipment.

subsea riser *n*: a vertical section of pipe that connects pipeline on the sea bottom to a production platform on the surface.

substructure *n*: the foundation on which the derrick or mast and usually the drawworks sit.

subsurface *adj*: below the surface of the earth (e.g., subsurface rocks).

subsurface safety valve *n*: see *tubing safety valve*.

supply reel *n*: a spool that holds drilling line.

surface casing *n*: see *surface pipe*.

surface hole *n*: that part of the wellbore that is drilled below the conductor hole but above the intermediate hole.

surface pipe *n*: the first string of casing (after the conductor pipe) that is set in a well. It varies in length from a few hundred to several thousand feet (metres). Some states require a minimum length to protect freshwater sands. Compare *conductor casing*.

surface safety valve *n*: a valve, mounted in the Christmas tree assembly, that stops the flow of fluids from the well if damage occurs to the assembly.

surface stack *n*: a blowout preventer stack mounted on top of the casing string at or near the surface of the ground or the water.

swamp barge *n*: see *inland barge rig*.

swamper *n*: (slang) a helper on a truck, tractor, or other machine.

sweet crude *n*: see *sweet crude oil*.

sweet crude oil *n*: oil containing little or no sulfur, especially little or no hydrogen sulfide.

swivel *n*: a rotary tool that is hung from the rotary hook and the traveling block to suspend the drill stem and to permit it to rotate freely. It also provides a connection for the rotary hose and a passageway for the flow of drilling fluid into the drill stem.

swivel stem *n*: a length of pipe inside the swivel that is installed to the swivel's washpipe and to which the kelly (or a kelly accessory, such as the upper kelly cock) is attached. It conducts drilling mud from the washpipe and to the drill stem. See *washpipe*.

T

tally *v*: to measure and record the total length of pipe, casing, or tubing that is to be run in a well.

tapered bowl *n*: a fitting, usually divided into two halves, that crew members place inside the master bushing to hold the slips.

TD *abbr*: total depth.

tear down *v*: see *rig down.*

telescoping joint *n*: a device used in the marine riser system of a mobile offshore drilling rig to compensate for the vertical motion of the rig caused by wind, waves, or weather.

tensioner system *n*: a system of devices installed on a floating offshore drilling rig to maintain a constant tension on the riser pipe, despite any vertical motion made by the rig.

tension-leg platform rig *n*: a compliant offshore drilling platform used to drill development wells. See *platform rig.*

thermally stable polycrystalline diamond bit *n*: a special type of fixed-head bit that has synthetic diamond cutters that do not disintegrate at high temperatures. Compare *polycrystalline diamond compact.*

thribble *n*: a stand of pipe made up of three joints and handled as a unit. Compare *single, double, fourble.*

thribble board *n*: the name used for the derrickhand's working platform, the monkeyboard, when it is located at a height in the derrick equal to three lengths of pipe joined together. Compare *double board, fourble board.*

throw the chain *v*: to flip the spinning chain up from a tool joint box so that the chain wraps around the tool joint pin after it is stabbed into the box. The stand or joint of drill pipe is turned or spun by a pull on the spinning chain from the cathead on the drawworks.

thruster *n*: see *dynamic positioning.*

tight formation *n*: a petroleum- or water-bearing formation of relatively low porosity and permeability.

tight hole *n*: 1. a well about which information is restricted for security or competitive reasons. 2. a section of the hole that, for some reason, is undergauge. For example, a bit that is worn undergauge will drill a tight hole.

Tinkerbell line *n*: see *Geronimo.*

ton *n*: 1. (nautical) a volume measure equal to 100 square feet applied to mobile offshore drilling rigs. 2. a measure of weight equal to 2,000 pounds. 3. (metric) a measure of weight equal to 1,000 kilograms. Usually spelled "tonne."

tong dies *n pl*: very hard and brittle pieces of serrated steel that are installed in the tongs and that grip or bite into the tool joint of drill pipe when the tongs are latched onto the pipe.

tong hand *n*: the member of the drilling crew who handles the tongs.

tong pull line *n*: a length of wire rope one end of which is connected to the end of the tongs and the other end of which is connected to the automatic cathead on the drawworks. When the driller actuates the cathead, it takes in the tong line and exerts force on the tong to either make up or break out drill pipe.

tongs *n pl*: the large wrenches used to make up or break out drill pipe, casing, tubing, or other pipe; variously called casing tongs, pipe tongs, and so forth, according to the specific use.

ton-mile *n*: the unit of service given by a hoisting line in moving 1 ton of load over a distance of 1 mile.

tonne (t) *n*: a mass unit in the metric system equal to 1,000 kilograms.

tool joint *n*: a heavy coupling element for drill pipe.

toolpush *n*: Canadian term for toolpusher. See *toolpusher.*

toolpusher *n*: an employee of a drilling contractor who is in charge of the entire drilling crew and the drilling rig. Also called a drilling foreman, rig manager, rig superintendent, or rig supervisor.

top drive *n*: a device similar to a power swivel that is used in place of the rotary table to turn the drill stem. Hung from the hook of the traveling block, a top drive also suspends the drill stem in the hole and includes power tongs. Modern top drives combine elevators, tongs, swivel, and hook.

top-drive system *n*: see *top drive.*

top plug *n*: a cement wiper plug that follows cement slurry down the casing. It goes before the drilling fluid used to displace the cement from the casing and separates the fluid from the slurry. See *cementing, wiper plug.*

torque *n*: the turning force that is applied to a shaft or other rotary mechanism to cause it to rotate or tend to do so. Torque is measured in units of length and force (foot-pounds, newton-metres).

total depth (TD) *n*: the maximum depth reached in a well.

tour *n*: a working shift for drilling crew or other oilfield workers. On rigs where a tour is 8 hours, they are called daylight, afternoon (or evening), and morning. Sometimes 12-hour tours are used, especially on offshore rigs, where they are called simply day tour and night tour.

tower *n*: 1. a vertical vessel such as an absorber, fractionator, or still. 2. a cooling tower.

transmission *n*: the gear or chain arrangement by which power is transmitted from the prime mover to the drawworks, the mud pump, or the rotary table of a drilling rig.

trap *n*: a body of permeable oil-bearing rock surrounded or overlain by an impermeable barrier that prevents oil from escaping. The types of traps are structural, stratigraphic, or a combination of these.

traveling block *n*: an arrangement of pulleys, or sheaves, through which drilling line is reeved and which moves up and down in the derrick or mast. See *block*.

trip *n*: the operation of hoisting the drill stem from and returning it to the wellbore. *v*: shortened form of "make a trip."

trip in *v*: see *go in the hole*.

triple *n*: see *thribble*.

trip out *v*: see *come out of the hole*.

tripping *n*: the operation of hoisting the drill stem out of and returning it to the wellbore. See *make a trip*.

tubing *n*: relatively small-diameter pipe that is run into a well to serve as a conduit for the passage of oil and gas to the surface.

tubing safety valve *n*: a device installed in the tubing string of a producing well to shut in the flow of production if the flow exceeds a preset rate. Also called subsurface safety valve.

tubular goods *n pl*: any kind of pipe. Oilfield tubular goods include tubing, casing, drill pipe, and line pipe. Also called tubulars.

tungsten carbide *n*: a fine, very hard, gray crystalline powder, a compound of tungsten and carbon. This compound is bonded with cobalt or nickel in cemented carbide compositions and used for cutting tools, abrasives, and dies.

tungsten carbide bit *n*: a type of roller cone bit with inserts made of tungsten carbide. Also called tungsten carbide insert bit.

tungsten carbide insert bit *n*: see *tungsten carbide bit*.

turbodrill *n*: a downhole motor that rotates a bit by the action of the drilling mud on turbine blades built into the tool. Most often used in directional drilling.

turnkey contract *n*: a drilling contract that calls for the payment of a stipulated amount to the drilling contractor on completion of the well. In a turnkey contract, the contractor furnishes all material and labor and controls the entire drilling operation, independent of operator supervision. A turnkey contract does not, as a rule, include the completion of a well as a producer.

turntable *n*: see *rotary table*.

turn to the right *v*: on a rotary rig, to rotate the drill stem clockwise. When drilling ahead, the expression "on bottom and turning to the right" indicates that drilling is proceeding normally.

U

uncased hole *n*: see *open hole*.

undergauge bit *n*: a bit whose outside diameter is worn to the point at which it is smaller than it was when new.

undergauge hole *n*: that portion of a borehole drilled with an undergauge bit.

upper kelly cock *n*: a valve installed above the kelly that can be manually closed to protect the rotary hose from high pressure that may exist in the drill stem.

upset *n*: thickness forged to the end of a tubular (such as drill pipe) to give the end extra strength. *v*: to forge the ends of tubular products so that the pipe wall acquires extra thickness and strength near the end. Upsetting is usually performed to provide the thickness needed to form threads so that the tubular goods can be connected.

V

vacuum degasser *n*: a device in which gas-cut mud is degassed by the action of a vacuum inside a tank.

V-belt *n*: a belt with a trapezoidal cross section, made to run in sheaves, or pulleys, with grooves of corresponding shape.

V-door *n*: an opening at floor level in a side of a derrick or a mast. The V-door is opposite the drawworks and is used as an entry to bring in drill pipe, casing, and other tools from the pipe rack.

voids *n pl*: cavities in a rock that do not contain solid material but may contain fluids.

vug *n*: 1. a cavity in a rock. 2. a small cavern, larger than a pore but too small to contain a person. Typically found in limestone subject to groundwater leaching.

W

waiting on cement (WOC) *adj*: pertaining to the time when drilling or completion operations are suspended so that the cement in a well can harden sufficiently.

water well *n*: a well drilled to obtain a fresh water supply to support drilling.

weight indicator *n*: an instrument near the driller's position on a drilling rig that shows both the weight of the drill stem that is hanging from the hook (hook load) and the weight that is placed on the bit by the drill collars (weight on bit).

weight on bit (WOB) *n*: the amount of downward force placed on the bit by the weight of the drill collars.

well *n*: the hole made by the drilling bit, which can be open, cased, or both. Also called borehole, hole, or wellbore.

wellbore *n*: a borehole; the hole drilled by the bit. Also called a borehole or hole.

well completion *n*: 1. the activities and methods of preparing a well for the production of oil and gas or for other purposes, such as injection; the method by which one or more flow paths for hydrocarbons are established between the reservoir and the surface. 2. the system of tubulars, packers, and other tools installed beneath the wellhead in the production casing—that is, the tool assembly that provides the hydrocarbon flow path or paths.

well control *n*: the methods used to control a kick and prevent a well from blowing out. Such techniques include, but are not limited to, keeping the borehole completely filled with drilling mud of the proper weight or density during all operations, exercising reasonable care when tripping pipe out of the hole to prevent swabbing, and keeping careful track of the amount of mud put into the hole to replace the volume of pipe removed from the hole during a trip.

wellhead *n*: the equipment installed at the surface of the wellbore.

well log *n*: see *log*.

well logging *n*: the recording of information about subsurface geologic formations, including records kept by the driller and records of mud and cutting analyses, core analysis, drill stem tests, and electric, acoustic, and nuclear procedures.

well site *n*: the place where a well is drilled.

wildcat *n*: a well drilled in an area where no oil or gas production exists.

wildcatter *n*: one who drills wildcat wells.

wiper plug *n*: a rubber-bodied, plastic- or aluminum-cored device used to separate cement and drilling fluid as they are being pumped down the inside of the casing during cementing operations. A wiper plug also removes drilling mud that adheres to the inside of the casing.

wireline *n*: a small-diameter metal line used in wireline operations. Also called slick line.

wireline operations *n pl*: the lowering of mechanical tools, such as valves and fishing tools, into the well for various purposes. Electric wireline operations, such as electric well logging and perforating, involve the use of conductor line, which in the oil patch is commonly but erroneously called wireline.

wire rope *n*: a cable composed of steel wires twisted around a central core of fiber or steel wire to create a rope of great strength and considerable flexibility. Often called cable or wireline; however, wireline is a single, slender metal rod, usually very flexible.

WOB *abbr*: weight on bit.

WOC *abbr*: waiting on cement; used in drilling reports.